new stars for old

STORIES FROM THE HISTORY OF ASTRONOMY

Marc Read

NEW STARS FOR OLD

FIRST EDITION
- 2013 -

Published by
Candy Jar Books
113-116 Bute Street,
Cardiff Bay, CF10 5EQ
www.candyjarbooks.co.uk

A catalogue record of this book is available
From the British Library

ISBN: 978-0-9571548-6-5

Printed and bound in the UK by
CPI Group (UK) Ltd, Croydon, CR0 4YY

Dedicated to my family, friends, and students
(past and present)

Contents

Introduction

Science is done by people.

That might seem a staggeringly obvious thing to say, but it's all too easy to forget. We're very used to the idea that when we think about art, music, literature, or architecture, we should learn a bit about the person who created the work, and about the society he or she lived in. Why not science, too?

Partly it's because we learn science in school with very little (if any) history involved. We're taught current theories, because what's the point in learning about things that turned out to be wrong? There's a lot to be said for that argument; after all, school science is confusing enough without throwing in a bunch of old ideas and names as well.

On the other hand, we can get the impression that scientific theories are somehow revealed by a booming voice from Heaven, and are set in stone once they've been established. Nothing could be further from the truth!

Science is an intensely human activity. Any theory that you've encountered was produced by someone who had a personality and a life outside the lab, who was worried about paying the bills, who listened to music and read books; someone who was very much part of society. Scientists aren't disembodied brains, but people who can take all of those cultural influences and add their own insights and creativity, all while working within a logical and mathematical framework.

The other point about science is that it changes as time goes by. Old theories are constantly being replaced by new ones, and the reasons for this are many and complicated. It's not always as simple as "theory A explains things better than theory B," especially if theory B had a few hundred years of tradition behind it, or the support of some powerful people.

Both of these points are especially clear in the history of astronomy. Many of the characters involved were much larger than life, with extraordinary stories to be told about them. Their debates burned slowly, with ideas sometimes disappearing for centuries before re-surfacing – astronomers generally feel themselves to be part of a very long tradition, with time on their side.

New Stars for Old is an exploration of the way in which these human influences have shaped science. Over the course of the book, you'll meet many of the people who changed the way that we think about the universe. Some of them are famous, but others will almost certainly be new to you. I've stuck pretty closely to the facts and to what we know of the characters' personalities. If you're interested in which bits are my own invention, there are some historical notes at the end of each chapter, but these aren't crucial to understanding any of the stories.

Not all of the figures I've selected were scientists as such, but those who weren't played important parts in changing the way that science was seen by people around them. I've always had a soft spot for people who seem fated to end up in the footnotes of more famous figures' biographies.

Another big theme of the book is the way that science (or more particularly astronomy) relates to other areas of life. Two of these areas which appear in many of the individual stories are astrology and religion.

Astrology and astronomy often got lumped together in previous ages, and the words were used pretty much interchangeably as it tended to be the same people who studied them. Although nowadays there is a clear separation between them, it's clear that one of the main reasons for studying the movements of the stars and planets (astronomy) in the first place was to help make predictions about what will happen on Earth (astrology). Individual astronomers throughout history held widely differing views about how this link should work, or whether astrology was at all valuable.

The historical links between religion and science are also more complicated than many people believe. Over the past couple of hundred years, a particular fight between one sort of Biblical literalism and one sort of atheism held by prominent scientific figures has dominated any discussion of this topic.

However, 'religion' and 'science' are huge labels, and you'll see through these stories that even within the Roman Catholic Church (for example) there was a very wide range of opinions. Not only that, but the overall consensus changed from one generation to the next (and there have always been mavericks). In times when religion made up a much larger part of people's lives and thoughts, it's hardly surprising that it tended to dominate a lot of discussion and surface in places that seem surprising to us.

The influence of personality and culture on our lives and our work; the urge to progress, to improve on previous ages; the uneasy relationships between various different ways of thinking – we might think that these are very modern concerns. In fact, they're all essential parts of what makes us human. We have more in common with past ages than we realise.

A Short Timeline

4th century BC

The Greek philosopher/scientists Eudoxus and Callipus prepare the first mathematical description of how the planets move, based on spheres within spheres, with the Earth in the centre. The theory is refined and popularised by the greatest philosopher of their time, Aristotle.

AD 140s-150s

Claudius Ptolemy, a Greek-Egyptian, produces several books analysing planetary motions using sets of off-set circles, rotating with differing speeds around the Earth. His theory is widely adopted and remains the received wisdom for over a thousand years.

300s

The Roman Emperor, Constantine, converts to Christianity and introduces it as the Empire's official religion.

390-410s

Amongst the many teachers and editors/commentators of Ptolemy's work are two more Greek-Egyptians, Theon and his daughter Hypatia (perhaps the first great female scientist). They become embroiled in local political struggles with tragic consequences for Hypatia.

Early to Mid 400s

The western half of the Roman Empire becomes increasingly unstable and finally collapses from a mix of internal strife and invasions from outside. Astronomical research continues in the eastern half, based on Constantinople (also known as Byzantium) under a series of philosophers/scientists.

The Early Middle Ages (sometimes called The Dark Ages)

Scientific learning, especially in astronomy, flourishes in the Muslim world, notably in North Africa and Spain. Arabic translations of Greek writing are highly valued, and much original work is done. From the twelfth century onwards, this work starts to become available in Latin translations and prompts the formation of universities across Europe.

1054

Chinese astronomers observe a supernova explosion creating the Crab Nebula.

1070s

It is believed the Bayeux Tapestry is woven.

1088

The first European University is opened in Bologna, Italy.

1167

Oxford University opens.

1202

Leonardo of Pisa, also known as Fibonacci, publishes his book, *Liber Abaci*. This includes details of the Fibonacci sequence: add the last two numbers to get the next.

1277

The Bishop of Paris, Stephen Tempier, condemns 219 heretical texts in an attempt to control academic publications. The controversy stems from scholars challenging aspects of Catholic doctrine. Thomas Aquinas was a key figure in this movement – a Roman Catholic priest and philosopher who had twenty books condemned.

1300s

The Renaissance begins. Prosperity in Europe is heavily influenced by a series of famines, plagues and uprisings. General unrest causes problems across France and England. Oxford University gradually becomes the world-leader for scientific research whilst, in Paris, Royal patronage enables some scholars to shine.

1337

The Hundred Years War between England and France begins.

1370

King Charles V establishes standardised time in Paris, decreeing all Church bells ring with those of the palaces.

1381

The Peasants' Revolt begins in England. The peasantry challenged customary feudal practices.

1424

Al-Kashi introduces the value of pi.

1450s

Johannes Gutenberg invents the printing press, allowing mass production of books and helping to spread academic ideas. Astronomers realise that Ptolemy's theories can be questioned and begin to move beyond his work rather than simply refining the details.

1453
The Byzantine Empire falls to Ottoman armies.
The Hundred Years War ends as France reclaims its territory.

1473
Nicholas Copernicus is born. Growing up to be a church official, he presents the first fully fledged alternative to Ptolemy's Earth-centred system. His Sun-centred theory generates much controversy when publicised by one of his disciples.

1476
William Caxton brings the first printing press to England.

1492
Columbus' voyages to America reveal that there is much that the Ancient Greeks were unaware of, contributing to a growing feeling that Western Culture is about to surpass that of the Greeks and the Romans. This Renaissance is just as much scientific as it is artistic.

1517
Luther nails 95 theses to the doors of the Castle Church in Wittenberg, Germany, spurring on the Protestant Reformation.

1534
Luther completes his translations of the Bible into German from Ancient Greek, making the text more accessible.

1543
Nicholas Copernicus dies the same year his theoretical book, *On*

the Revolutions of the Celestial Spheres, is published.
Andreas Vesalius writes *De Humani Corporis Fabrica,* the first detailed and illustrated textbook of human anatomy.

1546

Tycho Brahe is born. Growing up to be an astronomer, he establishes the first scientific research institute and develops instruments to accurately measure the movements of celestial objects. He demonstrates that the motion of comets smash through fixed spheres in the heavens and continues to build instruments which help him create a catalogue of star positions. His work attempts to combine the theories of Copernicus and Ptolemy.

1564

Galileo Galilei is born. He is the first to use a telescope to study the heavens and investigates the physics of falling masses. He realises that other planets have their own moons, that the moon has mountains and that the sun rotates. These discoveries promote the Copernican theory, his dedication to which leads to a famous dispute with the highest levels of the Roman Catholic Church.

1571

Johannes Kepler is born. He later becomes one of Tycho's assistants. His painstaking analysis of his Tycho's data lead him to the conclusion that Ptolemy and Copernicus are both wrong: orbits aren't combinations of circles at all, they are elliptical.

1582

Christopher Clavius proposes reform of the system of leap years.

1600

Giordano Bruno is executed as a heretic, at least partly as a result of promoting the Copernican Sun-centred theory and subsequently threatening the Roman Catholic faith.

1601

Tycho Brahe dies and his assistant Johannes Kepler succeeds him as Imperial Mathematician.

1608

Hans Lippershey invents the refracting telescope.

1630

Johannes Kepler dies.

1642

Galileo dies whilst under house arrest, accused of heresy due to his support of the Copernican theory.

Sir Isaac Newton is born. He overturns theories that have been accepted since Aristotle's days. He shows that the planets move according to the same laws that apply to objects on Earth (gravitational force) and invents new mathematical methods to study these phenomena.

1647

Johannes Hevelius maps the surface of the moon.

1676

Ole Romer measures the speed of light.

1682

Edmond Halley proposes the periodicity of Halley's Comet.

1714

Gabriel Fahrenheit invents the mercury thermometer.

1727

Sir Isaac Newton dies.

1

Composition

327 BC

In which we meet Aristotle, the greatest thinker of ancient Athens and former student at Plato's Academy, discover his somewhat unusual household arrangements, and consider whether his theory of the Celestial Spheres has any connection with art, religion or love.

Herpyllis stood back to consider her Master's new haircut critically. In her opinion, it was too short for such a distinguished man, and would have been more suited to one of the yobs who haunted the Agora, hanging around the marketplace to see and be seen; but she hadn't been asked her opinion. She turned her head, and simply raised an eyebrow. Lydias, her own young maid, cocked her head to one side and pursed her thin lips. Herpyllis nodded agreement, then placed the scissors down and stepped forward to slip her pale hands through Master's silver-grey locks, such of them as were left, and tousle them artistically.

Aristotle reached up to touch her on the wrist.

"That's enough, honey. Has Pyrrhaeus laid out my jewellery?"

She had selected the particular combination of sparkling rings to match Master's new sky-blue robes herself, and doubtless he knew it. However, she helped to maintain the fiction that his elderly valet still had his uses, rather than spending his time with his feet up in the kitchen flirting with the scullery girls.

"Yes, my lord."

She nodded to young Lydias, who carried the tray over to him. Her hands slipped lower to caress his shoulders as he slipped on the rings. She couldn't disapprove of this: displays of wealth were never unfashionable, and he was one of the wealthiest men in Athens, perhaps in all of Greece.

"Are you sure that the children won't disturb us, angel?"

"I'm sure, my lord. Pythias is safely with her tutor, and I think that little Nicomachus is on a trip to a farm with his nurse. You know how he adores animals."

She smiled happily. Little Nicomachus was getting so big, talking excitedly about his tiny world. How did he get to be four so quickly, their little boy? The days are very long, as a friend had said, but the years are very short. She wondered if she should share that thought with Master sometime. That was normally a bad idea, as he'd only turn it into some sort of meditation on the nature of time. She leant forward and kissed the top of his head instead. Nico was now the same age that his big sister Pythias had been when Mistress died, she reflected, and instinctively murmured a prayer to Hera.

"What was that, dear?"

Luckily, before she had to explain, or lie – Master was at best ambivalent about the whole idea of prayer – there was a cough from the doorway. Olympios the steward was there; how did he move so silently for such a giant of a man? Years of practice, she supposed.

"Excuse me, my lord, girls, but your visitor is here. I've put him in the library, my lord."

Aristotle glanced helplessly at the scrolls spread out before him. He'd been reading one even during his haircut, and Herpyllis had had to stop suddenly several times as he nodded violently.

"He wasn't supposed to be here until after lunch! But Lysippos never was good at keeping appointments. Ah well, this will have to wait."

He gestured to his half-finished lecture notes.

"Pylly, honey, you come with me to greet our honoured, and very early, guest. Olympios, fetch some light refreshments. Goat's milk for me."

Herpyllis looked at the visitor with interest. Lysippos had aged a lot more than Master since she'd last seen him at the Court in Macedon, what, seven years ago? Although they were roughly the same age, the sculptor was balding, his remaining hair a dull grey in contrast to Master's elegant silver – and she noticed a walking-stick leaning near to his chair. She smiled a greeting, sliding silently away from Aristotle's side to sit on a low stool, her appointed place for this room.

The artist nodded, his expression as serious as ever. He wore a perpetually worried look, and Herpyllis wondered whether he was happy with his fame and the riches that had accompanied it. She had spent her life as a slave to powerful and wealthy families, and the idea that money couldn't buy happiness was a puzzling one. She cast her eyes down as she, in turn, became the object of scrutiny.

"It's good to see you again, my old friend. You'll excuse me if I don't get up, your chair is far too comfortable for an old man. Speaking as an artistic professional, I see that your Herpyllis has grown into a stunning young woman – you always did have very good taste in girls, and in boys come to that. I understand that belated congratulations on the birth of a son to the two of you are in order? You must be very pleased."

Herpyllis had never known how to read Lysippos. He was saying pretty things, as always, but his heart never seemed to be in it. Perhaps he only expressed his true emotions through his art, not through his voice. Another thought not to share with Master!

Aristotle smiled. He had waited until his Pylly, his slavegirl, maid, and lover, had settled before striding into the room to clasp Lysippos around the shoulders.

"Thank you, on all counts. Yes, I have two children now. Little Pythias is ten already, and I can't quite believe that. Time does odd things. I trust that you had a pleasant journey? Are your quarters satisfactory?"

"I came straight here – I'm sure they'll be fine. I'm impressed by your set-up here, a fine complex of buildings for a college, my compliments to the architect. But I'm impatient to start while there's good daylight. We can talk while I sketch, so long as you don't get too animated."

"Do I have time for a glass of milk, or do you need to start drawing before my features change?"

Now, Master was so incredibly easy to read, the good humour never far from his eyes, his heart, his voice. Olympios, who was wearing his best concerned stewardly expression, stepped forward and handed his owner some fresh milk, almost colliding with young Lydias. The slender young girl stepped back, and easily knelt beside the stool, bearing the embroidery that they were working on.

"And I hope you don't mind if I have Herpyllis, and her own maid, here. If I'm going to be sat still for hours, I'd like to look at something pretty."

Lysippos had brought a young man with him. This curly-haired youth, hovering behind the chair, handed a board, a lead-tipped stylus, and papyrus to the sculptor. Herpyllis scanned him from behind her lashes, careful to avoid his gaze. She set herself the challenge of working out their relationship. Was he apprentice, servant, slave, lover or a combination of these roles?

Aristotle sat down on the high stool near his desk. The steward had, in his satchel, the scrolls that Master had been reading from the bedroom and now he laid them carefully on the cluttered surface. The philosopher glanced down at them, then questioningly up at Lysippos.

"No, my friend. If you start reading, I'll only see the top of your head."

He had already started to sketch. Herpyllis craned her neck, but couldn't see the dull lines left by the metal pen on the crisp sheet. Art always seemed a very special sort of magic, almost like the way Master could take the dullest of lecture notes and breath life and passion into them when standing before his élite students. He cared so deeply about getting everything just right, surely the sign of a genius.

"Turn your head to the left, if you please. You're obviously in the middle of something – why don't you tell me? I miss our conversations from the Court."

What a glittering affair the Court had been! Old Philip and young Alexander, godlike in their arrogance and finery, with the greatest talents of the world gathered around them. Mistress had bought new clothes every week, as befitted a princess, and it had been Herpyllis' job to keep them in perfect condition. Sometimes she regretted the move to Athens, but Master was doing so well here, and their mansion really was a dream house. She was so proud when she went to the marketplace and heard the stallholders and the nobles alike talking of the latest news from the Lyceum. Everyone agreed that it was an even finer college than the old Academy! She glanced across at Lysippos' young man, wondering whether he would be studying here or staying with his master. He did have such beautiful eyes.

She bent her head down to her work again. An intricate decorated border would make this old blanket look a hundred times better. She could almost hear Master's amused query, *Have you changed its essence or merely its appearance?*

Aristotle sighed. "Where should I put my hands? In my lap, like this? Oh, it's astronomy. You know that maths isn't my thing."

"Then get one of your staff to handle it."

The old artist was working very slowly and carefully, thickening and defining the lines with long, slow methodical sweeps.

"I normally do, but Kallipos is away on one of his observation tours, and I've left it too late to ask anyone else. It's fine, I remember the basics, there are just some oddities I can't quite reconcile. I don't want to say anything *wrong* now, do I?"

"Still the same perfectionist. How many of your students do you think would notice?" Was he teasing, or serious? It was hard to tell, with that flat voice and dry delivery; but the curly-haired youth smiled broadly from behind Lysippos. Herpyllis suddenly decided that he must be an apprentice. His stance was confident, not servile. There was none of that odd chemistry that she observed between, say, Master and the young men of the Lyceum who had crushes (or more) on him.

"Maybe you're right. It's more a matter of principle. I define the course of instruction here, and how can I do that if I don't understand it all myself? Kallipos has left me his notes, here, and it's all coming back to me. We were both taught astronomy by Eudoxos, you know."

"Not sure I've heard of him – did he write anything popular?"

"Not really. He studied under Plato, ran the Academy whenever the Old Man was away. Chipped away for years at one of those impossible challenges that the Old Man used to throw out to drive us all crazy. The thing is, he got somewhere."

Herpyllis smiled softly as she switched the colour of the thread. "The Old Man" – Aristotle had loved his teacher dearly, and often recounted anecdotes over dinner of life with him. Nevertheless, he realised his predecessor's limitations, and had no time for his mysticism. When he had returned to Athens, there had been much discussion about teaching at the Academy – but Master had found the syllabus there far too narrow for educating society's future leaders. She raised her eyes to gaze at Aristotle, wondering how different things could have been if she'd never been selected by Mistress, if the family hadn't moved, if... if... She murmured her thanks to the Fates.

"The Old Man wondered aloud – could the motion of the planets, looping madly in the sky, going back on themselves, be explained by some combination of simple, regular movements? Probably just a bit of rhetoric, but Eudoxos was a brilliant mathematician. It took him years, but he did it. If you imagine that each planet – including for the moment the Sun and the Moon – is a light, stuck to a sphere, and that sphere rolls around, but it's inside some *other* spheres all rolling in different directions, you can build up almost any motion, d'you see?"

Lysippos nodded.

"In my art, the most complex shape can be reduced to arcs of regular curves combined in unexpected ways."

"Precisely! Eudoxos worked it all out. Twenty-seven spheres, all told."

Twenty-seven, thought Herpyllis. *When I'm twenty-seven, he'll be... sixty. An auspicious age. And little Nico will be seven and starting lessons with a tutor.* She turned her head to share the thought with her maid, and couldn't help noticing that Lysippos' apprentice hurriedly averted his gaze. *So many complex movements, all reduced to the basics. Oh yes, Master should speak to me about that sometime.*

Lysippos' eyes flicked constantly between his subject and his sketch. She could see his concentration clearly in his furrowed brow. There was a long silence, not an uncomfortable one. It seemed that the artist was processing what Aristotle had said, while still working. Master sipped his milk, Lydias was humming quietly next to her, the apprentice was still hiding his glances at her. Presumably, twenty-seven distinct spheres were turning around different axes. Everything proceeded according to the dictates of the gods.

Aristotle finished his drink, and placed it down on the desk.

"A perfect solution to an impossible problem. Except that Kallipos, bless him, has been taking precise measurements of planetary movements for the last few years, from different places, and has found that it is far from perfect. He's added another seven spheres to make the planets loop around in just the right way."

Lysippos nodded. He seemed to be shading now, filling a space with short, precise, cross-hatchings.

"Another perfectionist. Are you *sure* you gentlemen don't believe in Plato's ideal forms?"

Aristotle chuckled.

"I'll explain both systems in my lecture, of course. I have two concerns, though."

The apprentice had stepped behind his master, to lean over and view the sketch. His look of rapt admiration thrilled Herpyllis. She knew what it was to devote her life to a person, a cause, a skill. She saw it in Master's face when he was lecturing, and she watched quietly from her seat by the door. Lysippos turned his head, sensing the presence of the young man. He smiled, the first genuine smile she had seen on his face that day. Maybe more than an apprentice after all?

"Maybe things are even more complicated than we currently think. Who knows what might be needed if Kallipos' latest observations don't tally? If so, then where will it end? Do we add more spheres every few years? Thirty-four could become forty."

Lysippos rested his stylus on the board, and looked straight at the philosopher.

"In my experience, my friend, things are always more complicated than they first appear. If you'll pardon my bluntness, that's why I could never be a system-builder, like you and Plato. I think that I have captured the very essence of my sitter – there, I thought you'd like that – on my papyrus, in my bronze. Then when everyone is admiring it and lauding it to the heavens, those perfectly spherical crystalline heavens of yours, I see the subject standing there with an expression on his face which is completely alien to my statue, and I know that once again, I have the wrong combination of basic elements. It's close enough to fool most observers, but that's not what you want, is it?"

Aristotle met his gaze.

"You speak well, my friend. While you're here, could I persuade you to give a lecture to my students on Truth in Art?"

"No, no. I've said my piece now, and spinning it to an hour's length would serve no purpose. But I'd willingly engage any students in discussion, if that would suit you. Now, could you turn your head a little to the right? Look just past your maid to the younger girl, if you please."

"Certainly. Like this?"

There was another long silence, broken only by Lydias' tuneless humming.

"All you can do is to teach the truth as you know it. Or does that sound too Socratic for your liking, my friend? Who knows? Perhaps this Kallipos will return and be able to confirm that thirty-four is exactly the right number of simple spheres to define and encircle the gods."

Aristotle rolled his eyes.

"Let's leave the gods out of this, hmm? It's the other concern of mine that's making this so difficult."

"Go on."

Herpyllis finished filling a particularly ornate key pattern, and snapped the thread. She always marvelled at the intricate designs that humans could produce, and saw no reason that the gods couldn't be equally complicated in their workings. That was an unfair jibe from the artist, who should know better.

"I'm caught in a dilemma, Lysippos. Maybe these spheres are just... mathematical tricks. They allow us to predict the planetary motions, but by accident and connivance rather than because they truly describe the structure of the world. In which case, I couldn't possibly lend them my authority – and yes, that does count, even beyond Athens. It's not enough to be right, your arguments have to be valid as well."

She had heard this many times. How could that apply to us mortals here below the unchanging heavens? Surely we're creatures of error and imperfection. What if my reasons for loving Master aren't what he would call valid ones? What matters is that I do love him. A sigh escaped her as she started the next interlocking part of the pattern, but nobody seemed to notice. Apart from the apprentice. Who, she supposed, had *valid arguments* to suppose that a slavegirl in his host's house was a proper object for his attentions, but who happened to be wrong. It cut both ways.

"And what's the other horn of your dilemma?"

"If the spheres are real, aethereal spheres – then why don't the spheres governing Mercury get muddled up with the spheres governing Venus? It won't do. The space between each pair of planets must be filled with counter-rotating spheres to make sure that each one starts from scratch, with no motion from the inner planets, d'you see? So now we'd need fifty-five spheres."

Lysippos had returned to sketching.

"Fifty-five is better or worse than thirty-four or twenty-seven, in your opinion? Surely you don't believe in lucky numbers. If it's fifty-five, then it's fifty-five."

The apprentice had produced another, smaller, board and

stylus and was also sketching. He grinned as Herpyllis noticed him, then raised an eyebrow. She nodded, flattered. Would she ever be cast in bronze for the lobby of the Lyceum, she wondered? She who had kept Master going through the despair following Mistress' sudden death? If things had been different... but they weren't, they couldn't be. Life is just as complicated as it needs to be. If it's fifty-five, then it's fifty-five. She decided that she liked Lysippos, after all.

Aristotle stretched.

"I can't sit here much longer, I'm too used to walking. Let me show you around – we can resume the sketch later. You're right, my friend, the truth is the important thing, no matter what. Fifty-five has a certain ring to it. I just wish I had the chance to talk it over with Kallipos before the lecture. Well, we seldom get everything we want."

He grinned and rose, then turned to his steward.

"Olympios, lunch will be at the fifth hour, in the blue room. Pylly, why don't you show Lysippos' assistant to his quarters?"

And as the young man offered his arm to help Herpyllis up, she wondered just how complicated it was to keep spheres safely separated.

Notes on Chapter One

Aristotle

Fictional characters:

Lydias (although we know that Herpyllis had her own maid)

Aristotle's household arrangements were as described. Herpyllis, his deceased wife's young slavegirl, became his constant companion. In his will, he freed her and left her money, slaves and property, commenting on how good she had been to him. The other slaves mentioned are named in his will. The philosopher was known for his fashionable appearance.

Lysippus was the most famous sculptor of the day, and a beautiful bust of Aristotle is attributed to him. (I have generally used the Greek "-os" ending for names, rather than the Latin "-us" except when the characters are very well-known and this would prove confusing. However, the Latin spelling is normally the more common one.) They worked together at the Macedonian Court at Pella, where Aristotle was the tutor of Alexander the Great.

Aristotle's astronomical writings were a tiny part of his work. He was a universal genius, commenting on every aspect of the natural and human world. If his books seem dry to us today (compared to those of his teacher, Plato) it is almost certainly because what have survived were his lecture-notes rather than anything more complete. He had a reputation for being an energetic and engaging teacher.

2

Evidence

AD 153

In which we jump across the Mediterranean to Egypt, visit Alexandria (founded by Greeks but now a part of the Roman Empire) and find that a mixture of religious piety, mathematical brilliance and crafty business sense lay behind the extraordinarily complicated planetary system of Claudius—Klaudios Ptolemy—Ptolemaeus—Ptolemaios.

"Shut up! I can hear you, I'm coming! Just keep quiet, for Bacchus' sake!"

Bacchus was to blame for all this, thought Klaudios as he rolled out of bed and into an old, threadbare, robe. His greying hair was dishevelled, his beard a little on the wild side, but what did such an eminent scholar care for such things? No, his attention was fully occupied with his thumping head and dry throat. He was getting too old for this sort of thing. He really should have known that a civilised dinner party to honour the new Director of the Library, the greatest collection of texts in the known world, would last into the small hours and involve wine, women and furious academic debate. Oh Athena, what had he said to the Head of Research at the Museum? It might be, in theory, a Temple of the Muses; but in practice, Klaudios knew that his standing there relied as much on who his friends were as on either intelligence or piety.

There was another hammering at his bedroom door. A quick glance through the window – the sun was over the palm-trees outside Hermes' temple already, it must be nearly noon. Damn, damn, he had arranged a mid-day meeting with Marcus Flavius the astrologer, newly arrived from Rome, hadn't he? Why hadn't his slaves woken him earlier? Useless, the lot of them.

He flung open the door, trying to get his thoughts together as his butler cowered before him.

"Is he here yet? He is? Damn, damn... show him to the study, get me clothes, and cold water. Lots of water. He can wait, he was the one who wanted to see me."

Some fifteen minutes later, he presented a slightly neater figure as the butler opened the study door for him, although his hair was still noticeably damp. His headache had subsided to a dull thudding, thank the gods, and he felt ready for anything that a mere commercial astrologer, even one with offices in the Temple district, could throw at him. This Marcus had only appeared a month or so ago, but was apparently poaching clients from the more established businesses left, right and centre. He'd probably just come to point out an error in the *Handy Tables*, some impossible conjunction or other, or Mercury getting too far away from the Sun. Then Klaudios could show him the master copy, and explain that it was a simple copying error – it always was. You just couldn't get the scribes these days, not for complicated pages of numerals. At least he'd be back in bed in, what, half an hour?

His visitor was surprisingly young – maybe mid-twenties – and very well-polished, even unctuous. He wore expensive silks, of midnight blue with vaguely astrological symbols tastelessly embroidered around the hem. Either business genuinely was booming, or he wanted to project that image. Klaudios sized him up quickly; probably he had a client base of well-to-do women of a certain age. Marcus put down the scroll he had been perusing, standing from the low stool, and smiled a very false smile.

"It's an honour to meet you, Claudius Ptolemaeus," he murmured in Latin.

Klaudios had precisely no time for that. Alexandria might be a melting-pot, and his taxes might end up with the Emperor, but his loyalties were firmly with the previous wave of conquerors and he felt Greek to the core. He shambled behind his desk, making a sweeping wave that could mean almost anything.

"What do you want, lad? I'm busy."

Was that a smirk on Marcus' clean-shaven face? Had he drawn his own conclusions already? Astrologers were good at that sort of thing, and Klaudios should know – he'd been one for years before taking up technical mathematics, and they were still his biggest audience. Lazy, oily, types who couldn't do the maths, who just wanted the results. No wonder his *Handy Tables* were outselling the *Syntaxis* by about twenty copies to one. Nobody these days seemed to be buying his handbook on astrological theory, either. Strange how the market shifted, he thought to himself, as he gathered up the voluminous folds of his clothing and reclined in as much comfort as he could manage. He vaguely wondered how he had acquired so many bruises on his legs last night.

Marcus spread his hands, a conciliatory gesture, head slightly bowed in deference, switching effortlessly to fluent if accented Greek.

"Well, sir, I'm puzzled. There's a discrepancy..."

"Hah! I knew it! You just can't get the scribes these days, you know? When I was a lad, they had values, training, even in long strings of numbers. How do you think we learned?"

"Indeed, sir. I don't think it's scribal error, as I've already compared my copies with those of several colleagues, and with the text in the Library."

Klaudios' eyes widened. This whippersnapper had connections

at the Library? Either he was a scholar, which made him potentially a colleague but more likely a rival, or... perhaps he'd caught the eye of one of the Library hierarchs. Or both?

"That's interesting, then. I'm sorry if I seemed a little short. It's just..."

He trailed off, waving vaguely, trying to will the demons in his head to stop their antics with a whispered prayer to Asklepios to heal him quickly. Marcus' easy smile was infuriating.

"Of course, sir, I understand completely. Well. Recently I've been trying to work through your theory of the Moon, sir, and I can't quite get things to work out right."

Klaudios swung himself upright, quicker than his body was expecting judging by the sharp pains that lanced across his forehead. He scowled them away, and kept the scowl on his face as he rested his elbows on the desk. He had been dreading this day, but he had always thought that when the blow came, it would have been from one of his colleagues, or someone new at Athens.

"The Moon. What's wrong with the Moon?"

"Well, sir, first of all let me say how wonderful your books are. We couldn't do our jobs without them, and you're obviously –"

"Yes, divinely inspired or a genius and you count yourself lucky to be alive in the same time as me and I've reformed the subject beyond all recognition. I know. Stick to the Moon."

"It's probably nothing. Your *Tables* tell us exactly where it will be, when it will rise, every day with miraculous accuracy. I should be happy with that, that's what my friends told me when I said I was coming to see you. But I've always been curious, sir, and I wanted to follow the theory for myself, so I've been working my way through the *Syntaxis*. It's eye-wateringly hard, and my maths isn't quite up to it, but some of the scholars up at the Museum have been helping me. I've had them check my working, but they can't find anything, so I've turned to you. You see –"

Klaudios couldn't stand it any more. If it all had to fall apart, let him be the one to pull out the key-stone.

"I know. The distances are all wrong. My explanation gets the positions right, but they predict that the Moon –"

His head slumped into his hand. After all, he couldn't say it. Let this rising star have his moment of glory, even if the blaze was from burning the *Syntaxis*.

Marcus was nodding along, and picked up the thought.

"The Moon, as seen from the Earth, should double in size every month as it approaches the Earth. Except that... well..."

He shrugged and raised an eyebrow, hands spread, the comic actor's stance to indicate surprise and disbelief.

"The mistake isn't in your mathematics, young man. The thing is, I can't see where I went wrong, either. As you so generously observed, the mechanism gets the position right every time. Sit down, lad. I've had this speech prepared for quite some time, so you might as well be the one to hear it."

Klaudios felt very old, very suddenly. Was it time to hand on the torch? A torch that continued to give light although its core was rotten? He'd intermittently rehearsed his excuses on behalf of the Moon ever since he had put together the original theory, but hadn't done so recently. He'd certainly never expected to be hungover when it came to the crunch.

"The Moon's different from the other planets, you see. I spent years working out just how everything up there moves – circles, circles, everywhere. I was eating, drinking, dreaming circles. I was consumed with circles – of course, it has to be circles, doesn't it? The heavens are unchanging and perfect, we all know that, and a circle is the only perfect shape, yes? If you've worked through the *Syntaxis*, you'll know that it's not just Earth-centred circles, that doesn't get it right. The Ancients were wrong. You've got to have circles on circles, but that isn't enough – the circles speed up, they slow down, their centres are all different, it's crazy up there but it works. You wouldn't expect the gods to be simple, would you? The priests will tell you that only the initiates get to know the true stories, and what's astronomy but the priesthood of the heavens?"

Marcus frowned. He'd heard that old Ptolemy was a little bit off-balance; to borrow a word from the *Syntaxis*, "eccentric". He'd certainly not been warned about this religious slant. Astrology was just a job to him, a good job, paying well and with decent social status, and he was good at it. Wasn't the mathematical side of things just the same?

Klaudios leaned forward.

"That's why if you've any sense you'll get out of your current game, my boy. Why sell trinkets to the temple-goers when you could be a priest instead?"

Was a reply required? The young man shrugged mentally, and tried to prod this increasingly unstable prodigy back into the right orbit.

"But the Moon, sir, you were talking about the Moon!"

"So I was. The gods know that I struggled long and hard with the other planets. Mercury and Venus, now, they're the real trouble. You'd think that they would wander away from the Sun, but they don't, and it took me three whole years just to fix their cycles, their epicycles, their eccentrics and their equant points... but I did it, with Athena's help. So I thought that the Moon would be easy. Everyone knows how the Moon moves, don't they? We've seen it from childhood, before we even know about the other planets. Except it doesn't work. I couldn't get the mechanism to work.

"I was in despair, I didn't eat or sleep, I could barely get through the days. Until I had a vision of one final circle, the missing piece! You stare, lad, but I really mean a vision, sent from the gods. What if the epicentre itself cycled around the Earth? Crazy, I know, but the Moon is the Triple Goddess, the realm of magic and mystery, so why shouldn't it have a magical, mysterious, motion, an extra wheel in the machinery?"

He lifted his head, eyes shining, the headache completely forgotten.

"And it worked! It worked! I was the first person, with the gods' help, to crack it – the position of the Moon, perfectly predictable after all."

Marcus coughed softly. He felt rather brutal interrupting the old man, but this conversation was going precisely nowhere.

"Except, sir, it didn't. Work, that is. The distances. The Moon doesn't change its size, does it? It waxes and wanes, but the disk is always about the size of a sestertius piece held at arm's length."

"Except it didn't. The distances. Yes, yes. Can you even begin to imagine how I felt?"

Perhaps, after all, he could. Even though he hadn't seen as much, done as much, as this latter-day Aristotle, Marcus had known the weight of crushing disappointment enough times in his life to get just an inkling of the emotions that Claudius – no, Klaudios – must have experienced. He lowered his head, folded his hands in his lap, carefully adopted his most calming, conciliatory voice. He hadn't come to needle the old chap, he had genuinely thought that he had misunderstood the text, had miscalculated. Everyone knew that *Ptolemy* didn't get things wrong! He was a household name, not just in Alexandria, but even in Athens, in Rome itself.

"So, sir, what did you do? And what can we do?"

He was implicated, too. Had nobody else noticed this before, could he really have been the first to spot the inconsistency? Was that a matter for pride, or for regret?

"Do? A good, practical question, from a good practical businessman."

Klaudios poured two cups of cold water from the jug his butler had thoughtfully provided, sipping one and gesturing for his visitor to take the other.

"I'm not proud of what I did. I didn't lie – it would be blasphemous to lie in such a holy task."

His conscience pricked him – or was that the sudden shock of the cold water? – and he promised himself that one of these days, he would silently edit his astrology manual to remove all that nonsense about finding his ideas in previously-undiscovered Babylonian scrolls. But that was what the market wanted to hear! Back to the matter at hand.

"I didn't exactly draw attention to the problem, either. I knew that people who read the *Syntaxis* would either be mathematicians who didn't know the astronomy, or astronomers who didn't understand the maths."

He looked up from his cup, pleadingly. "Was that so wrong?"

Marcus shook his head slowly. He was glad he had the cup as a prop to play with now, and busied himself in sipping the water. It saved him from having to say anything else.

"As for what we can do now, I've been thinking of playing the same trick. Did you know that I'm writing another book, explaining how all the systems in the *Syntaxis* fit together? You didn't? Here, have a look at my *Hypotheses*, such as they are."

He pushed a loose sheet of papyrus, covered with his latest work, across the desk.

Marcus scanned the page quickly.

"Oh. I see."

As chance would have it, Klaudios had been working on the average distances of the planets. Or was it chance? Perhaps they were both being swept along in some elaborate joke being played by the gods on those who would penetrate their heavenly realm. He shook his head briskly. Some of this religious mania was beginning to rub off on him, and he didn't like how it felt.

"I'm going to leave it at the average distances, I'm not going to show my workings. My old teachers would have a fit! Because..."

"Because if you showed the workings, sir, it would be obvious to every fool who can read the *Handy Tables* that the Moon isn't doing what it should. I see."

Klaudios put down his cup. He started to speak, but visibly halted to consider before resuming.

"Marcus, you're the first person to have spotted this. I hope you'll be the last, at least until after I'm long gone. Can I have your oath that you'll not mention it to any of your friends, to anyone else at all? When they ask about our meeting, tell them that the great Ptolemy showed you the error in your maths. If my reputation for accuracy goes, then so does everything else. Can I have your oath?"

Marcus felt a surge of pride at this confirmation of his own abilities, his insight. He sipped slowly, turning the matter over. He had nothing to gain from the destruction of this fellow's reputation, nothing whatever, and Ptolemy would be the best possible ally in getting his business up and running. A personal endorsement from the Prince of Astronomy above the door to his offices. Appearances together at public events... it was easy enough to decide.

"You have my oath. Shall I swear it formally before a priest?"

"No, man, the gods will witness it here well enough. I think you'd better go, now. I have a lot to think about."

Marcus placed the cup down delicately next to the jug, pushed the papyrus back across the desk, and stood. He turned to leave, without interrupting Klaudios any further; the latter was sunk deep in thought.

As his foot crossed the threshold, it came to him.

"Sir! I have it! I have the answer!"

Klaudios' face was a picture, thought Marcus. At least three competing emotions took it in turns to govern his features, in very rapid succession.

"What do you mean? You barely understand the mathematics!"

Marcus felt his confidence welling. Human weakness and frailty, the limits of reason; these were his professional meat and drink. Admittedly, not on such a universal scale.

"What if it displeases the gods for us mortals to know their secrets? No, hear me out. We know that our reasoning is correct, but our eyesight, our senses, our poor, feeble, human senses... how easy it is to mislead them, to judge things wrongly. Can you, Klaudios, judge accurately how far away a horseman near the horizon might be? Are our mariners never led astray in their navigation? Not from reason, but from the weakness of our senses when looking at objects far away. Indeed, the further away the object is, the less likely we are to perceive its size and distance correctly..."

He paused, as understanding dawned on the astronomer's face.

"And our senses cannot therefore be relied upon to judge sizes when the distances involved are so huge, such as..."

"That between the Earth and the Moon."

The philosophers stared at each other. Could that be it? The universe was governed by reason, but here on this corruptible and corrupted Earth, we cannot see properly?

Klaudios grabbed papyrus and ink, knocking over the jar in his hurry. The cold, clear water trickled across the desk, pooling on a long-abandoned scroll containing scores of detailed planetary observations from Thebes. It started to blotch and fade, unnoticed, as the two men, the two colleagues, set to work, finding the right phrase to sell their new, improved, truth to the world. The breeze in the palm-trees laughed gently.

Notes on Chapter Two

Ptolemy

Fictional characters:
Marcus Flavius

Ptolemy was the greatest astronomer of the Classical world. We know very little about his personal life, other than that he started his career as what we would call an astrologer. We have to be cautious, as until the eighteenth century or so the words "astronomer," "astrologer" and "mathematician" were used completely interchangeably. I have used them in their modern senses throughout. He certainly saw astronomy as a religious calling.

He was no relation to the Ptolemy Dynasty who ruled Egypt (being surnamed Windsor today doesn't make someone a member of the Royal Family). This confused historians in the Middle Ages, who often depicted our Ptolemy as an astronomer-king.

His astronomical system was incredibly complicated, as described in the chapter. The diagrams in history books with the planets moving in simple circles are a huge simplification. The Ptolemaic System could predict the positions of the planets with pin-point accuracy. However, it did predict that the moon should move in impossible ways. The compromise solution of skipping over this with a mention of the difficulty of judging distances was, indeed, the one that he adopted although there is no evidence that anyone else was involved. It is extraordinary that he got away with this cover-up until the Renaissance (when, for the first time, astronomers appeared who were even better mathematicians).

3

Passion

AD 391

In which an enigmatic, brilliant and controversial astronomical heroine, Hypatia of Alexandria, is seen through the eyes of one of her students, and we discover that science, religion, love and politics can combine in many different ways.

The lodgings above Paulos' shop
The Street of the Bakers
Alexandria

To the most respected Theon, arch-mathematician, formerly of the Museum,

Greetings,

I write as your humble student of astronomy, Matthew son of James.

I was shocked to learn of the decree shutting the Museum and the Library as injurious to the Christian Faith. Please be assured that not all of the Christian community here support the actions of our nominal leader, the Patriarch, nor, if reports are to be believed, of his real pay-master the Emperor of Rome. I have learned much in my scant two years with you, and wish to pursue my studies further in whatever way possible. Is it true that you are establishing a Platonic Academy, similar to that in Athens itself?

Please let me know on what financial terms you will be accepting pupils and I shall be glad to enter into any contract that is required, with my parents as guarantors. I hope that this may turn out to be a new opportunity for you to establish yourself as an independent figure and to pursue that scholarly career which has made your name known across the civilised world.

With my humblest devotion, Matthew James

AD 393

The Platonic Academy
Opposite the smaller Temple of Osiris
Alexandria

To my beloved little brother, Joshua,

I hope that this letter finds you well, Josh. Thanks for your last – yes, it's a pity that I haven't been home for so long, but everything here is so busy! I've finally mastered the arithmetic and geometry that I need to move on to more advanced courses.

I think you'd love it here at the Academy. The food isn't as good at home, and I'm convinced that our wine is watered to the point that its Essential Substance has changed, but it's all so exciting. Theon – we call him The Boss – is completing his new commentary on Ptolemy's *Handy Tables*, and he's picked his favourite students to act as scribes since the commercial ones are uniformly awful. They just can't concentrate on rows of numbers. I bet Ptolemy never had this problem! So guess what? He's picked me! It's a real honour, and I feel especially proud because we Christians are in such a minority here. And I'm the youngest of the bunch, too…the others are all in their twenties.

So Big Brother is doing very well. Reassure Mum and Dad that things are great, and their money is being well spent. The Boss reckons I could be a professional mathematician, maybe even a philosopher. I'll write to them when I get a chance, but you know how Dad insists that our letters be word perfect. You like my little subjunctive there? That's the sort of thing he'd appreciate. Give little Mary a kiss from me, too.

Love and peace in the Lord, Matt

AD 394

The Platonic Academy
Opposite the smaller Temple of Osiris
Alexandria

To my dear little sister Mary,

Hi, honey! I hope that Josh or Mum or Dad will help you with any tricky words in this letter. I miss you and it was so nice to see you last week at your birthday celebrations. I'm sorry that I'm always at the Academy, but I've got such a lot of work to do. I'm sure you want me to work hard, just like your tutor tells you!

Do you remember that you said you didn't want to grow up, because women never get to do interesting things? I have to write to let you know that you're wrong! I've just started the course here in Modern Philosophy and my tutor is a woman – Mary, you would love her! She's called Hypatia, and she's the daughter of Theon, our headmaster. She's really, really clever, and everyone says that she helps Theon to write his famous books. She's also really, really pretty, and she's only a few years older than me, so you see you can still do all sorts of fun and useful things if you keep on with your lessons.

Love and hugs and a kiss from Matty

AD 397

The Academy
Alexandria

To Mistress Hypatia,

Greetings from your humble student, Matthew. Please forgive my presumption in writing to you rather than seeking to talk to you after our classes. I have to write this down in a letter because otherwise I won't remember, or be able to communicate, everything that I want to say to you. Even after these years, I'm still tongue-tied around you in front of others. I know that it's due to base human emotions from the lowest part of the tripartite soul, but I'm not strong enough to overcome them, not strong like you.

Where do you draw that strength from? You, and your father, have taught me well that in the pursuit of the truth, we must follow the questions and the arguments wherever they lead, like great Plato and Socrates before him. So I observe that you are single, although you are beautiful, clever, and strong-willed. Can it merely be because of the latter? Surely there are men enough who would love you for that, a true love such as we find in the *Symposium*? You rebuff any such talk, but it is a valid question and could throw light on the nature of our humanity. Do you suppress the feelings arising from our corrupt and sinful nature, or simply not feel them in the first place?

On a more pressing point, I've been reading the copy of Alcinous' *Handbook* which you lent me, and it's interesting stuff although I can see why you don't use it in class. It's so contradictory and confusing. Who knows what he was thinking of? In three separate places, he refers to astronomy as being a mathematical art, and a theological art, and a physical art. Which is just to say that it's an art, and he doesn't even seem too sure of that. I'm losing sleep over this one, Mistress, because I know that he was a true disciple of Plato, albeit many generations removed. Did he err through sin, or sin through error? I must know! It's hard to believe that he was writing at the same time as Ptolemy. What a philosopher we missed there! It doesn't do to bury oneself in mathematics and ignore religious and spiritual questions, does it?

Now that I've raised the points, can we discuss them sometime? Maybe over a glass of wine and a meal at the Olive Grove taverna? Although we are immortal souls first and foremost, surely we owe it to our fleshly bodies to satisfy at least some of their needs.

Your student and servant, Matthew James

AD 399

The Academy
Alexandria

Most honoured Theon,

It is with joy in my heart that I write to confirm what we agreed verbally last night. I would be delighted to join your family of teachers, by becoming Assistant Instructor in advanced geometry and technical astronomy. I confirm that I will be guided by your ever-esteemed daughter in establishing the curriculum and in selecting the students who have attained the required standards for such a demanding course of studies. I look forward with a keen spirit to working more closely with her, and with you.

Your humblest colleague, Matthew James

AD 402

The Academy
Alexandria

Hypatia, my dear,

I have no words for your loss. It might seem too trite for me to say it, but Theon genuinely was like a father to me these last few years. Working late into the night, poring over the trigonometry and the geometry, just the three of us – I can't believe that I'll never again hear his soft voice, puzzle over his illegible handwriting. Since my own family moved away to Carthage, the two of you, and my work at the Academy and in the Church, have been my entire life. I don't know how I'll come out of this.

You're being so strong about it, so (dare I say it) stoical. But grief for your father is not only a natural human emotion, it's commended to us both by my religion and by yours. I'm always here for you. I know that you've lived your life in the World of the Forms, with your head full of ideas that I can only dream about. Now you find yourself in a public position, as Head of the Academy, and you might find it useful to have someone a little more worldly around, if only to handle the day-to-day business affairs. Just ask, my dear, and I'm your man. Now and always. Would you object if I prayed for your father's soul according to my own rite?

With heart-felt sympathy,

Your nearly-brother, M.

AD 403

The Teachers' Chambers
The Platonic Academy
Alexandria

Dear Joshua,

How's business? I understand that you're more or less running the show now, and that Dad's taking less of an active part. I'm really pleased for you. You were always the one with the flair for trade, my head is stuck in the clouds, or above them. You must come and visit sometime soon. Alexandria would be a great place to show the children as soon as they're old enough to appreciate it.

I'm doing well, better than I ever expected. You know that I'm Deputy Head of the Academy now, Hypatia's old job? And Hypatia is my boss, and my best friend. She's like a sister to me, Josh. I don't know whether you'd get on with her, she's a bit of an acquired taste. She's stunning, with those big almond-shaped eyes, her perfect skin, her long, lustrous hair... you're going to say that I sound like a young man who's lost his head over some girl.

That's exactly what I am. I can't stop thinking about her. She's everything that I've ever wanted in anyone, pretty, scarily intelligent, educated, strong-willed, spiritual. She'd laugh to hear herself called a saint, but she is!

If you met her, you'd think that she's cold, unemotional, logic personified. That's the face she presents to the world, but when you really know her, she's got a wonderful wry sense of humour, and a killing turn of phrase. I think that now old Theon's gone, I'm the only person who really knows her in all the world. Part of her act is to pretend that she has no feelings, romantic included – I think it's easier for a woman to get on in this world if she behaves like that. (You've probably heard about the incident some years ago, when one of the students asked her for a date, probably as a stupid dare. She pulled out her bloody menstrual cloth, and demanded, "How can anyone love this?" I thought that was wonderful!)

There's the obvious trouble that she's pretty much past child-bearing age, but that's only one of the reasons to get married, isn't it? It's better to marry than to burn, and I'm certainly burning now. The deeper problem is that in these last few years, we've genuinely grown to think of each other as brother and sister, and she's very comfortable with that relationship. I've never dared to push it further, but my obsession is growing to the point where I find it difficult to concentrate on my work (I almost had Mercury wandering too far from the Sun without noticing the other day!) You've always had luck with women, Josh – I know it's meant to be the other way around, but could you share some words of advice with your big brother, or at least wish him luck in what might be a doomed venture? I know she thinks a lot of me, but this is uncharted territory for both of us.

The old crowd from St Luke's send their best wishes, by the way. I'm deacon there now, and that takes up more time than I would like. Still, it's all good honest work.

Love to Xanthe and the kids, Matt

AD 406

The Platonic Academy
The Grove of Academe, Athens

My dearest Hypatia,

For so I think of you now and will do for ever, while the stars continue in their courses. I love you. You know that I love you. Don't worry, I'm not writing to try to persuade you to change your mind. We all have our destined paths. You are my Earth, the centre of my cosmos, the source of all order. I am doomed to orbit you forever, approaching and retreating from you in turn, your Moon, with my face always turned towards you. The whole intricate system dances to a tune that you can hear, have always heard, which I only dimly perceive when I strain my senses to the utmost.

I am as well as I can be after your rebuff. No, that's the wrong word, there was nothing cruel in what you did, you were merely stating the unalterable laws which you can see in everything. Your fate is to remain alone, seemingly aloof, and mine is – what? What can you see in my future, in the influences of the wandering planets on my unimportant trajectory through the heavens?

I am writing now, after my self-imposed silence, to let you know that I have moved again from Thebes and am now in Athens. Plutarch – the head of the Academy here – has been kind to me and given me a post teaching Modern Philosophy. I strive every day to be as good a teacher as you were, even if some of the texts still confuse me. Should you ever wish to contact me, your letters would reach me there. If you do not write, I shall endeavour to understand and to accept, if not to approve. Once I thought that my approval would please you.

With all the devotion of my heart, your Matthew

AD 410

The Priest's House
Next to the Church of St John the Divine
Athens

Dear Mary,

Who can foresee the path which God has chosen for us? I agree that a few years ago, I would have sworn blindly that a career in Academic philosophy and astronomy was exactly right for me. I've changed a lot since then. In many ways I've grown up, learned to be self-reliant and prayerful in a way that I never truly was before. Then I saw as through a glass, darkly – you know the rest.

I suppose that it wasn't until I had to teach philosophy, rather than study it, that I had really considered its implications. I'm not going to renounce my previous views or repent my scholarly path to this point (isn't the goal more important than the road?) Far too much is tied up with them, both intellectually and emotionally. It simply became increasingly uncomfortable for me to be just about the only Christian on the teaching staff, and I had to consider where my priorities lay, and lie. Trust me when I say that there was a huge amount at stake for me, in terms of how I think about myself and what I've done with my life, and how I've acted to people. I'm not going to hate myself for the sins of my youth, either of commission or omission. Have you read Augustine of Hippo? He has a lot that's valuable about this sort of thing.

Since my ordination, I've spent a lot of time re-thinking my actions up to now, and I realise that I've been pretty selfish and not a very good brother either to you or to Josh. Now that you've a family of your own on the way, I hope that you can find it within you to accept the money enclosed with this note. It represents some of the savings that I maintained for many years, hoping that certain things would change and that maybe I'd get married. It's a long story, but one that I'll tell you when we next meet up. Some things are best left for a conversation, however convenient letters can be.

Do come and visit when you can. The priest's house here has a lovely view out over the olive groves.

Love and peace in the Lord, Matty

AD 415

The Priest's House
Next to the Church of St John the Divine
Athens

To the most respected Orestes, Prefect of Egypt,

Thank you for the letter and packet, which have reached me safely. I no longer work at the Academy, but they were forwarded to me here at St John's.

I am frankly in shock at the news you report. I studied under Hypatia and later worked with her for several years at the Alexandrian Academy, and can't quite believe that she has departed this life, and in such a manner. She was always so thoughtful, even if at times she could be angry, and I can't possibly imagine what she could have said or done to have provoked people quite so much. I admit that she could also be stubborn and her tenacity could infuriate people, but even so this is quite extraordinary. I hope that you can write further with more details, as your letter raises far more questions than it answers.

How are you, as the direct representative of the Roman Empire, involved? You say that she was deliberately and cruelly murdered in the events surrounding a near-riot, and that local Christian political militants have claimed responsibility – who, what, why?

I have, or at least had, some contacts in that community and if I can assist your investigation or the process of justice in any way at all, I would be ready to return to my home-town at a moment's notice.

Could you also tell me – did she ever marry, does she leave a family? If so, I would be delighted to contribute to help support them on a regular basis. She was a dear colleague to me and the Academic world will mourn her. At least her true memorial will endure as long as the heavens continue to turn in their circles. For generations to come until the ending of the world, where Ptolemy is remembered then surely so will Theon and Hypatia.

I can also tell you that the packet arrived in excellent condition, a copy of Alcinous' *Handbook of Philosophy* with annotations and commentary in her own hand. Do you know if these incisive and compelling observations on this little-known text are also to be found in the Academy library? If not, I should be delighted to copy it for that purpose.

Your humble servant, Father Matthew of St John

Notes on Chapter Three

Hypatia

Fictional characters:
Matthew, Joshua, Mary

The story of Hypatia, daughter of Theon and head of the Alexandrian Academy, has had a large impact in modern times. As a female academic in a position of public power, who remained single all her life (the incident with the menstrual cloth is related by several biographers) and who was murdered by Christians, she has inspired much writing, some painting and even a film.

Her murder was, almost certainly, unrelated to her personal religious beliefs or teachings. Alexandria was a melting-pot city torn by racial and religious tensions, which often boiled over into riots. Hypatia was an outspoken supporter of the Roman Prefect, Orestes, in his struggles against the Church – but for political rather than religious reasons. She numbered several Christians amongst her students, and although I have invented Matthew and his family, his story is plausible.

Unfortunately, her contribution to astronomy itself is hard to assess. Her surviving work was written jointly with her father, consisting of commentaries on Ptolemy and updated editions of his astronomical tables.

4

Growth

AD 426

In which we follow the life of Proclus from his adolescence as a rich law student, to his death as an ascetic and influential philosopher of religion, touching on the way in which he wove together many strands of intellectual life to make a harmonious whole.

Dear Diary,

Father says that I have progressed enough with my lessons to go to law school, at last. Thank the gods! I have sweated blood over my Classical Greek, my Latin, my thrice-accursed mathematics and astronomy, my rhetoric and my gymnastics. I know that I have to excel at everything if I am to live up to my family name, but it has been so hard.

The other boys are all in awe of my father. I can't tell whether it's because of his talent, his position or his sheer wealth. Then I think, in ten years I will be running the law firm in his place, and young men who are mere toddlers today will be admiring me in the same way. I try not to be arrogant or proud, but it is so very hard when we are clearly much higher-born than the minor merchant families of this awful district. We can hold our heads up high even in the great city of Constantinople itself.

Mother says that the law school in Alexandria is the most respected in the world, but that I shall have to learn the various Egyptian languages as well if I am to be a part of the city's civic life. I don't know whether I want to be. What is Egypt good for but supplying corn? I shall plunder knowledge like a pirate, and bring it back in triumph to the Empire.

There are a million and one things to do before I leave. I must pack, and say farewell to all of my extended family. I don't think that I'll bother sending messages to my so-called friends here; I'll never be returning to this awful suburban existence. My destiny lies in the highest positions in the Empire, in Constantinople; I can't think why Father doesn't want his family with him next to the Courts.

What an odd thought that my next entry will be written in Alexandria! You, oh my diary, are far too important to be left loose on the boat and will be packed away in the hold with my dearest possessions.

AD 428

So, as I sit here on the quayside, the gulls wheeling and turning endlessly, what do I feel about Egypt? My head is crammed almost to bursting with the laws of city-states I didn't even know existed, written by petty despots and tyrants that have almost faded from history. This is a country that always looks towards the past in order to prepare for, and cope with, the future. Is that romantic and valuable, or foolish in this world in which the new is always arising from the old? Too much philosophy rots a man's mind, prevents him from acting, from becoming that which he is destined to be. How unlike these gulls, who have no separation between desire and action.

Am I a man, though? In simple legal terms, I am now old enough to take on my civic duties in the Empire, and to assist my father in Court. Surely, that is enough. I am looking forward and planning my life in terms of years, rather than days. Surely, that means that I am ready for more than a mere apprenticeship? Can two years of rote learning, of tests, of recitations, count for so little?

I suspect that having to pick up new languages so quickly was better for my education than all the tables of trials and all the scrolls of great legal speeches. Adaptability will be the key to success in this world. I can imagine the voice of my philosophy tutor: *Is success in this world really your aim, young Proklos? Then I hope that you achieve precisely that.*

But then again, he's a philosophy tutor, and I'm on my way to become a lawyer in the greatest city that the world has ever seen. Rather puts things in perspective, doesn't it?

AD 429

The case went as well as could have been expected. Honestly, clients are often their own worst enemies – it's just as well that they have lawyers to look out for their best interests. I've always thought that tax evasion is a particularly insidious crime. If individuals don't pull their weight and contribute to the Empire in due proportion to their wealth, how can they complain when they see cutbacks in Imperial projects? At the moment, the army is going underfunded, and quite frankly I'm worried when we see what inroads the barbarians have been making on our territories year after year.

I've been thinking a lot about the notion of civic duty today, surrounded as I am by all the pomp of the Imperial Capital. Do we also owe wider duties to the gods, to use the abilities which they have given us? My friends tease me that I'm far too bright to waste my life defending amoral merchants, and I'll admit that some of the less savoury characters represented by the firm give one pause for thought. Helen thinks that I ought to be a philosopher or a priest, and dedicate my life to the service of The Truth and the gods. Joanna is trying to push me towards Christianity, but there are too many paradoxes there for me. I know that I shouldn't be swayed by the opinions of girls, for all sorts of reasons, but Helen is very persuasive, and father doesn't really need so many assistants to keep the business going. If I chose a different path to his, would he be angry or proud? I wish we had talked more when I was younger, but he was never there.

AD 430

Athena, how could I have been so blind? I spent nearly two years here, and I thought that what I was doing was studying. Now, on my return to Egpyt, I see that I had never really opened my eyes throughout that stay. I thought that law was difficult and useful, and that language was simply a tool for communicating intentions. I knew intellectually that laws are all made by mortals, but until this week I hadn't thought it through. The universe around us was made by immortals, and therefore is infinitely more worthy of study.

Forgive me, Athena, for having laughed at your servants. For even those who do not know it, even the Christians and Jews and atheists among them, honour you by their words, by their lives. The scholars of the immortal world can be called many things – philosophers, mathematicians, mystics, astronomers, theologians, priests – but the labels don't matter, do they? I laughed because I thought that they were simply playing games with words, and because I had only really encountered Aristotle's works. I know now that I have been guilty of deprecating a whole city, because I found its most famous temple ugly.

Why did I have to come so far from home to encounter the writings of the philosopher Iamblichos and his ancient masters, Socrates and Plato? If these men could combine all the titles I have listed above, then cannot another in our modern age? I am proud, yes, and Helen would call me arrogant and immodest, but how can one search for the truth if one does not acknowledge the truth about oneself?

I know why I had to come here – to remove myself from distractions. Yes, including Helen. And father, and mother, and the law firm, and everything that goes with that life. Am I far from home? That depends entirely on what I think 'home' means to me now. I am at a crossroads in my life. Would a sacrifice at Athena's temple be impious, or a good start? I must find someone at the Academy here who can understand me.

AD 431

I am convinced that places have attendant spirits. How else can I explain the extraordinary sense of homecoming, of recognition, that swept over me when we landed at the Piraeos, and I walked the road to Athens itself? I needed no guide. I believe that I was recalling memories from the time before my soul had a mortal body, and this must be mediated by the daemons that haunt this most holy of cities. The city of my long-dead masters, and of the Goddess herself.

I am fully aware that the Academy here is not actually in the Grove of Academe any more. When I have a chance to explore a little more, I shall certainly pay my respects and pour a libation on the ruins of Father Plato's original buildings. But the Academic idea is embodied in people as well as places, and Syrianos strikes me as a worthy successor to the Ancients themselves. I have been warmly welcomed here, and feel that I am truly at home for perhaps the first time. To think that I fell so in love with Alexandria! Now that the first flames of passion have faded, I see that it was a childish infatuation with surface appearances. The essence of truth does not lie in books, and to go beyond and behind the books some combination of knowledge, spirit, and conversation is required. All of those are here.

A few years ago I would have felt aggrieved that all my studies count for nothing now, and that I am being treated the same as any other young new arrival. I repent of my foolishness. I had been trying to run in the Olympic Games when I was barely fit for preliminary training, and seeing Syrianos and his colleagues makes me realise just how mistaken and pretentious I had been. My new life starts now, and I have sworn to obey my teachers faithfully and to live according to the precepts of the scholar. But did my new life *have* to start with basic mathematics and astronomy?

AD 440

Highest Athena, hear the prayer of your faithful servant, Proklos of the Academy.

Tomorrow, I am to give my inaugural lecture as Professor of Philosophy and I am dedicating it, and my subsequent researches here, to your honour and glory. I am nervous, my Goddess, so nervous to follow in the footsteps of your great teachers and adherents. Watch over my faltering steps and voice, and shield me from the spite and envy of my enemies. Cover me with your aegis, and let me care only for your voice, which I ever hear in my dreams and my meditation.

I shall honour you in my speech tomorrow through a new approach to my subject, which will take me years to develop in full. I have not yet shared this with any mortal, and await your blessing in any vision you vouchsafe to me.

The Ancients, my predecessors here, divided philosophical learning into theology as the study of the Divine, mathematics as the study of the unchanging, and physics as the study of this mortal realm.

What do we say to the astronomers? They study the unchanging motions of the heavenly bodies: but when they *explain* these motions, they use notions proper to physics, they talk of mechanisms and rolling spheres and real circles. Are they polluting the heavens, or elevating earthly notions?

Oh Athena, you have granted me a glimpse of the truth. Astronomy is composed truly of both natures. It is the proper subject to study that which mingles earth and heaven, mortal and divine. It is, it must be, the key to studying the human soul! In the motions of the planets, our own inner workings are revealed.

All praise be to the Olympian gods. May I always speak in their name.

AD 450

What beautiful nightmares the astronomers have given us! Ptolemy's whirling circles, spinning around within Eudoxos' celestial spheres – or the equally complicated systems of Aristarchos – what are we to do with them? I am drawn to and repelled by their systems in equal measure, but I can no longer sit on the fence. As the Head of the Academy, the Platonic Scholarch, I am being pressed on all sides for a reform of our astronomical syllabus.

It is clear to me that the movements of the planets are whole and entire of themselves, not merely combinations of "simple" motions. Who are we to judge what is simple to the gods? The tables of Ptolemy, Theon, Hypatia serve to predict the positions of the heavenly bodies, but can we truthfully call their geometrical methods an *explanation* of what is happening? Introducing equants, eccentrics and all the rest of the technical apparatus so beloved of our astronomers today raises many more questions than it can possibly answer. It rips Heaven itself into tiny fragments, each behaving like a toy-maker's fever-dream. No, the realm of the gods cannot be divided like that.

Does that mean that we abandon explanation, or treat our mathematical constructions as mere fictions? Surely not! The challenge is to find explanations of the right sort. Every physical phenomenon has an underlying metaphysical cause. We must find the true causes of heavenly movements. A division into spheres and circles is right and fitting as displaying the perfection and eternity of the whole; what offends the gods is the thought that our whirling epicycles can correspond to the Unity that underlies the many-formed world of our perception.

In pedagogical terms, of course our astronomers must teach the methods which allow us to use planetary motions for practical purposes. They must not, however, mislead their students into believing that the underlying causes, for surely there must be such, correspond precisely to what we have so far concocted.

AD 453

It is something of a shock to be back in Athens after my tour of neighbouring states and countries, and to plunge straight into the concerns of the Academy. While everything has been running nicely during my sabbatical, all the decisions that are so important to my staff (appointments, promotions, curriculum review, pay) have been left for my return. I suppose that I should be flattered that my opinions are considered so crucial.

Such a change from the last year, when I had to consider only myself and my relation to the gods. (If not to the Christian god: I am relieved beyond measure that some of the simmering tensions within the city have died down, and wonder if my absence played a role in that.) How different are the perceptions of humans around the civilised world!

My mind, my body and my soul are still reeling from the myriad experiences. My studies of religious practices and of theology were noble and worthy, but the mind alone cannot prepare the soul for the full impact of actual mystical contact with the One who is manifested in so many forms. There are multiple paths, leading to the same Truth, and now I have trodden all of the roads that I found open to me. The Mystery Cults of the Olympians are all so very different, but all so very similar in their essentials.

The students have given me the affectionate nick-name of The Universal Priest. I think this is true in a way they did not intend: if I am a priest (and what remains of my legal mind tells me that having been initiated, I am by definition), I count myself as a Priest of the Universe. Simplicity expressed through Complexity leading back to Simplicity.

And we all need a connection with the underlying Unity more than ever in these troubled times. I have not yet been able to digest, to process, the news that Constantinople herself was attacked by the Huns, and her defence was not through arms or valour but by a tribute of gold. Maybe the petty merchants and lawyers of my youth proved to be doing their duty more than I ever believed.

AD 476

So Rome has finally fallen to the barbarians. By all accounts, they have looted and burned much of the city: for the past few months we have welcomed as our guests various scholars who found it convenient to visit Athens, and it is likely that they will now join the Academy or the other schools here formally.

How much store can we set by the fate of one city, one clan, one family, one man? The Undivided and Eternal Empire of Augustus fragmented, recombined, fell into civil war and changed its shape, its laws, its customs, its language over time. Those of us left, in the increasingly isolated Eastern remains of that proud dominion, would do well to remember that nothing beneath the lunar spheres is permanent. We live in the world of Becoming, not in the world of Being to which we aspire.

Some of my brethren are calling this a judgment of the gods, or of fate. But fate is blind, and the motivations of the gods are beyond our comprehension. The Christians are being as loud and unreasonable as ever: it seems that each of their ridiculous bishops has a different theory about why Rome had to fall.

All we can do is to act as god-fearing citizens should act, with a clear duty towards all of our fellows.

Do the barbarians also worship our gods, under some of their myriad names? Do their priests preach to them their duty to glorify their own temples and rites? All my instincts tell me that the riddle can only be solved by meeting them and learning their secret rituals, but my colleagues here are adamant that it is too dangerous for, well, an old man. Some Universal Priest I have turned out to be! However, I act not only for myself but for my School, and I take my oaths seriously.

My visions have increased in number and clarity. Increasingly it is Ares who visits me in my sleep. The elders of Athens ask me for my advice on matters of state, and my head is filled with thoughts of war. Is it time to hand over the burden of leadership, or would that be shirking my duty?

AD 484

I, Proklos, Scholarch of the Academy of Athens, Successor to Plato, do hereby declare and swear that I have no known living relatives.

I own no immovable property or land, living purely within the official residence provided by the Academy.

When I am dead, I direct that all of my slaves be freed and that my other property be sold, and that the wealth be transferred to the funds of the Academy to be used as the new Scholarch shall see fit. I ask him as a friend to remember those who have been kind to me, and those temples which have nourished my soul and helped my studies, but I understand that more pressing needs could arise.

I do not believe that I owe any debts, whether financial or emotional, to any other person.

Let me be buried in some spot sacred to my beloved Athena. Any memorial is entirely at the discretion of the Academy.

Let it be said of me that I did my duty.

In the name of the gods I swear this will.

Notes on Chapter Four

Proclus/Proklos

Fictional characters:
Helena, Joanna

Proclus is one of the "Neo-Platonists." They took the writings of Plato and infused them with spiritual and mystical significance. There was a long-running strand of Greek thought, seen most clearly in Pythagoras, claiming that mathematics was the underlying mystical explanation for the nature of the world. This elevated its study almost to a religion. Proclus took this to extremes, producing a heady synthesis of philosophy, physics, mathematics, mysticism, astrology and astronomy which reminds us of similar New Age movements today. Other Neo-Platonists had attempted this, but Proclus was also a very competent mathematician and could actually do the mathematical astronomy that they only talked about.

His surprising path from arrogant young lawyer to ascetic – but sociable – holy man and teacher, as recounted in the chapter, is confirmed by several ancient authors. He did undertake a sabbatical tour of the Eastern Mediterranean investigating religious cults, and experienced increasingly frequent visions of Athena (in one case, rescuing a statue of her from a militant Christian group, claiming to be acting on her explicit instructions).

His surviving work is unfortunately very hard to unravel, as he wrote in a pithy and awkward style. This has led to considerable debate amongst historians and philosophers. There is a lot of confusion over what he believed to be real, and what fictional, in astronomy but the position he holds in the chapter fits well with his writings.

5

Conjunction

AD 1142

In which we learn of another Norman Conquest, and see that Greek ideas lived on in the Muslim world during the fragmentation of European civilisation, only to be reintroduced to the West through a happy alignment of trade, geography and politics.

The Room above Luca's Fish Shop, by the Harbour
Palermo
Sicily

Dear Father Geoffrey,

It has been many months since I left the Abbey at Winchester, and I have often wished to find the time to send you a letter. How well I remember your sermons, the only source of light and heat in the damp abbey! I hope that you are well, and that the local villagers are starting to appreciate the benefits of your attempts to reform their system of government. Young William of Caen, who is here on his youthful travels, tells me that he will be visiting relatives near to you shortly, and so I am taking advantages of his good offices to send you this overdue report on my progress here. I remain eternally grateful to you for the education you gave me, and your encouragement of my interest in matters of natural philosophy and astronomy, and hope that all is well with you and with my former brothers. Please allow them to read any of this message which you see fit to share and which would edify or amuse them.

You will doubtless be aware that "the Kingdom of Sicily" is something of a misnomer. Under our current King Roger II, our realm extends well beyond the island, extending up the Italian Peninsula to the borders of Rome itself. It is a wondrous land, filled with sun-drenched plains, precipitous mountains, and foods and drinks unknown in the cold, dark North. The wine is strong and revives the heart, and the spices from our Eastern trade are certainly nothing like the good solid English fare of the old refectory.

My heart must always belong to England (I owe her the duty of a son to his parents, growing up amidst her gentle rain and drinking her ale). But our forefathers made two great Conquests in the past hundred years, and the sweet land of Sicily has very many advantages. Given its situation at the heart of everything and the incredible things that are happening here, I have little doubt that Roger the Conqueror will be remembered when his distant cousin William is merely a note in a chronicle. What we have achieved here is little short of miraculous, and I apologise if I seem carried away. I had no dreams that such a place could exist on the Earth; I had been taught that everything here below the Moon was corrupt and decayed, with perfection only attainable in the higher heavens. I encourage you to visit if you possibly can! Yes, we won it by force of arms, but we are building it into the greatest agent of God's purpose that one could imagine.

Sicily itself has been improved beyond measure by our architecture. I understand that the towns were poor and mean before we introduced centralised taxation and a strong feudal system, but now they flourish. It was a shock at first to see a good, solid, motte-and-bailey castle topped by a dome (yes, a dome), or to see our new churches built with arches and towers that seem to have wandered from nearby mosques, but now it seems a perfect symbol of the reality of the kingdom; that we Normans have taken the best traditions of all the nations surrounding us, and harnessed them to our own ends without losing our identity. My countrymen, my fellow subjects of the Sicilian crown, are true Normans in valour and daring, whether intellectual or martial, whatever their racial background, faith or mother tongue – and all appreciate that a strong government is required to save them from the banditry and piracy otherwise endemic to the Mediterranean climate. I think that there are some valuable lessons to be learned here for those of our kind who rule their baronies by pretending that only those of true Norman blood are real people, and that the native peasants are merely there to provide serfs (you know precisely the sort of people I mean).

My journey here was an eventful one, from the heartlands of Normandy through petty kingdoms to Rome following a well-trodden Pilgrims' Path. As a young scholar there was plenty to engage my attention. Although I still have some regrets at having left the Order, my newfound freedoms allow me to worship in different and less formal ways. I know that is not that path that you have chosen, but I have had my new vocation confirmed many times since crossing the Channel. I have seen and experienced things which would be simply impossible for a regular monastic (at least, one who took his vows seriously). I won't make you blush by recounting the odd things which befall a party of young pilgrims, but my initiation into the secular world was certainly rapid!

We moved south for weeks, always south, through increasingly arid regions as the Sun beat down, to the Kingdom itself and finally to Palermo. We passed through scenes of turmoil and war; Roger has been hard-pushed to define and defend his borders, and there have been ill-conceived rebellions and risings from the petty Italian nobles who do not relish learning good honest French. It seems that we have now reached a natural limit. The political situation with the various Popes (at least one of whom was briefly our prisoner) and the so-called Holy Roman Empire would make your head spin; it certainly leaves me thoroughly confused. From a strategic point of view, the current Northern border is held by our Counts and Dukes to be defensible (once duly fortified), and from the civic viewpoint the subjects seem genuinely pleased with the new common currency that is making trade between cities flourish. Maybe it won't be long before Normandy is using ducats, too, although I imagine that the English Saxons would object.

I must add that the climate is not congenial to us scholars (it is frequently hard to concentrate), and I pity the knights their mail coats in such unending warm weather. Another benefit of the secular life; I have swapped my woollen robes for a thin tunic. The local girls wear clothing that would lead to their rapid death from cold in Winchester, and I cannot find it in my heart to complain, although this also does nothing to help my concentration.

My work does, I truly believe, give due praise to the Lord albeit not in a form that the Abbey would recognise. You know that I travelled here to slake my thirst for languages, and now I am almost drowning in them. We are at a great crossroads of trade here. To the North, we have the Papal States and beyond that the Empire, bringing Latin, Italian, and German speakers (and their books). To the South we have the kingdoms and emirates of North Africa, bringing Arabic speakers (and their books). To the East we have the fading but still powerful Byzantine Empire, bringing Greek speakers (and their books). And to the West we have another mongrel state, the caliphates of Moorish Spain, bringing more Arabs, but also Spanish speakers (and their books) – and many Jews involved in cross-Mediterranean trading, although I have been disappointed that Hebrew is not commonly spoken amongst them, at least in the presence of gentiles (they prefer to use Arabic, something of a common language in this clime. Even our king is fluent in the language). Here in the court, we speak French but any competent merchant, priest or noble can be counted upon to speak at least two other languages.

You can imagine that it is a translator's dream. We scholars are like pigs in muck, revelling in the huge resources at our disposal. What is more, at Roger's court, the burghers and the guildsmen are all keen to learn whatever is useful and can advance their own individual causes, regardless of the source. I sometimes feel uncomfortable with the amount of religious freedom we allow here – mosques and synagogues are found along good Catholic churches, using both Western and Eastern rites, and the poor peasants who haven't had the education that I've enjoyed can seem confused about which way to turn. However, our priests (mostly from solid Norman stock) don't seem overly worried and I am willing to believe that they have the state's best interests at heart. Certainly it encourages trading, and while I might worry about the immortal souls of those I see in the street, where there is an exchange of commodities there is also an exchange of language, of ideas, and (as I've mentioned) of texts.

We're busy as bees, translating material from any of the languages I've mentioned into any of the others. The things that I have learned! Luckily, a large community of us is required. In an English town, one linguist might struggle to find work. Here we are essential support to each other, since many of the concepts we're dealing with are highly technical and require considerable explanation. I feel almost ashamed to have come from such a backwater, since nobody is interested in studying the Saxon or Gaelic which I acquired so painstakingly, and I'll confess that their tales of monsters and heroes, of battles and kings, stirring though they undoubtedly are, pale compared to the pure nectar of Homer and the Ancients. One of these days, if I can find the time, I'd love to produce a good sound French version, or maybe even Saxon.

We Latinists get all the glory, however, since translation is normally *into* rather than *from* our preferred tongue and our market is potentially huge. Which monastery, which palace, which Imperial Library, would want to be without the latest Latin editions of the works from Saracen Africa, or Persia? It has been extraordinary to me to see how advanced their ideas are, how much they have understood of God's creation even if they fail to understand His purpose. I can barely imagine what the union of this pagan philosophy with true Christian thinking will bring forth. Have you heard of the new type of school called a "University," recently established in Northern Italy, which is attempting exactly such a combination? The idea is to unite scholars from different backgrounds, with different interests, and to teach every one of them the basics of every art and science known to us and to the Ancients, before they decide how to specialise. Although it is a radical idea, it might lead to great things; we translators are trying to persuade King Roger that a similar institution here would increase his prestige and consolidate his intellectual reputation.

I have only recently learned how little we in the North know, by contrast to the stream of ideas that flows from the Ancient Greeks. The modern Greeks and Byzantines have kept certain philosophical traditions flourishing, but most of the explication and transmission of material has come to us from the West and the South. I have fallen in love with astronomy, all over again, as surely it allows us insight into the realm of God Himself and His angels. The perfection of a planet's movement is itself a hymn. Back in the monastery, I counted myself learned for knowing the basics of eclipses and recognising the seasonal changes of the constellations. Now I know that I thought as a child. Hundreds of years before the Christ was born, pagan philosophers had discovered the secret workings of the heavens, and have refined their work in an unbroken chain ever since. The names of Aristotle and Ptolemy will perhaps be new to you, but they saw further and deeper than many of our own theologians.

Unfortunately, I have not yet been able to acquire the full writings of the great theorists – one Aristarchus apparently developed a theory that the Sun is at the centre of the universe, and I would need to see his arguments with my own eyes before I could assess them. Similarly, Ptolemy produced a work of great subtlety known today as *The Almagest* (which signifies 'greatest' in Arabic) but copies seem to be rare, and only available in those cities with great astronomical traditions. The Muslims have a chain of observatories stretching throughout their whole Empire where trained men do nothing but measure the positions of the stars and the planets. However, I have managed to gather the basics of their systems from their commentators. My current project is to translate the works of one such scholar, whose name may be turned letter-by-letter into our script as Ibn Bajjah. I have Latinised this as Avempace. He is an old man now (some say he is dead), living (if he does) in Fez, a poet and a doctor as well as an astronomer. His works are especially respected by the Spanish Jews, which is how they have reached me, rather than directly – I told you that the circumstances here favour such lucky connections.

Within any tradition there will be differences of opinion, as you will doubtless be aware. Astronomers from the earliest times have been divided as to whether the planets are attached to concentric spheres, as taught by Aristotle, or to off-centre combinations of circles, as maintained by Ptolemy. When such great minds differ, what can we humbler intellects add to the debate? Luckily, my Avempace has been able to provide a compromise which seems just and fitting to both parties, namely off-centred spheres which can in their intricate combinations move the planets around, making them now approach the Earth and now recede towards highest Heaven. The spheres of Aristotle could not explain why the planets vary in brightness, but are certainly to be preferred to the extraordinarily complicated patterns demanded by Ptolemy. One of my Byzantine colleagues tells of a modern Greek tradition wherein both circles and spheres are considered to be merely imaginary calculating devices, on the authority of Proclus, Simplicius and other such obscure authors, but once again I cannot lay my hands on the texts to establish the true facts, and cannot bring myself to believe that such elegant examples of God's creative handiwork could in fact be nothing but phantasms of the human mind. I must travel further East when my work here is done – there are more languages to be learned, and so many more authorities to be recovered in the field of astronomy alone.

Do not worry, though – the study of the superlunary world has not obsessed me completely, and, besides, to be appreciated in all of its complexity requires the mastery of more mathematics than I currently possess. You will be pleased that chronicles, histories, and travellers' tales of far-off lands abound, and all of them are vouched for by merchants, scholars and soldiers who have been there themselves. The world is a very much larger place than we Normans have been accustomed to think, both the physical world and that of the mind. When our forefathers in Scandinavia told their sagas, bound by a world of ice, blood and gold, they imagined that this Middle-Earth between heaven and hell was founded on glory, and the skalds sang suitably. Our Norman troubadours and minstrels celebrate a different, more civilised, age – but surely the true music of this world is the music of the Spheres, and we scholars are those that hear snatches of it down here below the moon and hum it to an often uncaring world. Those who recognise the tune can do so independently of their language or their religion, although only we Christians appreciate its true significance and I remain convinced that French is as good a tongue as any, and better than most, for the expression of complicated ideas.

I trust that all is well with you, and if you can find sufficient time and a suitable messenger I look forward to a reply. Memory is an odd thing, and sometimes when I dream, I consider myself still to be a part of your community. I hope that, despite the pain you felt when you released me from my oaths, you are proud of me in some way and can see that, beneath the appearances, we are not so very different. Do mention me to father if you see him.

Peace and love in the name of the Lord, Richard of Wykeham

Notes on Chapter Five

King Roger II and the Translators

Fictional characters:
Geoffrey, William of Caen, Richard of Wykeham

Most intellectual pursuits stopped in Europe in the long period after the final collapse of the Western Roman Empire. This was formerly known as the "Dark Ages" but many historians today prefer the "Early Middle Ages." However, work continued in the Muslim world, from Arabia, across North Africa, to Moorish Spain. The Muslim intellectual tradition arose from Greek works, often translated into Arabic by Christian scholars. Aristotle was revered by many Muslim philosophers – and attacked as being irreligious by many others. Mathematics and technical astronomy were popular areas of study, and many refinements of Ptolemy's theories were proposed.

There is much speculation by historians about why Europe suddenly exploded with academic activity, starting in the twelfth century, but two elements seem critical. The first was the translation of material from Arabic into Latin, and the (somewhat surprising) activities of Roger II, Norman King of Sicily, were hugely important early on in this movement. The second was the genuinely new idea of a *university*, where students had to study a little of everything ('the seven liberal arts') before specialising. This produced a pool of decision-makers, clergymen and royal advisers who were generalists, and those who stayed for further study at least knew the basics of general intellectual culture (including, of course, astronomy).

6

Hypothesis

AD 1277

In which the freedom of acadmic expression in the University of Paris collides with the authority of the Church, shaping the rollercoaster career of that unlikely hero, Thomas Aquinas, and the philosophical movement he created.

"It's a disaster!"

"Life will never be the same. They're cutting their own throats."

"How could they do this?"

"It's an outrage – I'll be leaving as soon as I can!"

Robert of Canterbury had heard the hubbub of a dozen over-excited scholars even through the thick door of his lodging-house, but on entering the common-room he was almost overwhelmed by the sheer volume made by his countrymen. This really wasn't the sort of thing one expected from a group of English academics, even those who had gone native. Robert himself was pretty much fresh from his Oxford degree, and was still getting used to the voluble and demonstrative Parisian locals on his way to and from the teaching rooms he hired. He was used to orderly, measured, debate. It was shocking to hear the other members of the English Hall expressing themselves so freely and (even worse) all talking at the same time.

Andrew the Scot, leaning back against the door frame, edged

to one side to let Robert past. At least the Scotsman seemed relatively calm, but nothing much had ever been known to move Andrew.

"Morning, Rob. You picked an interesting couple of days to take a break. The Bishop's gone mad, forced through the Condemnation. Want a hand with your bags?"

Robert nodded gratefully. He had over-packed hugely for a short trip, nominally to compare preaching styles, but really to observe agricultural methods over here; he couldn't quite shake off his own farming family history, however grand a scholar he was becoming. It seemed clear to him that, whatever the news, the other residents of his Hall were merely working themselves up into a frenzy with no actual exchange of information. He shook his head sadly as he reflected that these were some of the finest minds of his age, and here they were reacting like... peasants. Just the sort of thing he'd have expected from his Kentish cousins! He and Andrew struggled through the crowd (it seemed that some of the Germans were here too, and the common-room could barely hold them all). The noise abated somewhat as he climbed the stairs to his own small chamber. He smiled his thanks to Andrew.

"I thought that the Condemnation was dead in the water. Want to fill me in on the details?"

"I'd love to. There's been nobody speaking sense this morning around here. I only stayed for the entertainment value. Old William looks fit to explode! It's got them all so excited that they've even let the undergrads join in their discussions for a change, bless them."

The tall Scot smiled affectionately.

"It wouldn't be easy to talk here. Want to find a tavern? Somewhere where nobody will care about church politics, or understand our English if our thoughts should wander after the first few glasses?"

Robert ruefully thought that it was hard enough to understand Andrew's English even when the Scot was sober. He'd have more of a chance with Latin, even through that awful accent. It was even worse in other modern languages. He paused in unfastening his cloak, pinning it back up again against the perpetual drizzle.

"Sure! I'll get the first bottle and we can split the cost of any food?"

Money was a perpetual source of worry, and he liked to have the details arranged in advance without raising anyone's expectations. If whatever had happened damaged the credibility of Paris as a university, if his income from his students dried up, he'd simply have to risk his chances back home. As it was, the profit he made from teaching barely covered his own tuition fees and living expenses. Or maybe he could move somewhere warmer? No, it had to be Paris or Oxford, and Oxford was so far off the beaten Continental track that it didn't get the international students. Besides, who would teach him theology anywhere but here?

"Agreed. Coming?"

Andrew slipped back down the stairs and out through the commotion, the shorter, younger man following in his wake. Robert wondered whether his friends – no, colleagues – no, *friends* had even noticed him. The words he heard as he left were identical to those he had heard as he entered, although the accents and presumably the speakers had all shifted around.

As they settled into a dark corner, the rough red wine burning their throats pleasantly, Robert felt that he could relax.

"I suppose it's all still related to our jolly Italian friend, bless his soul?"

Andrew knocked back most of his wine, reaching for the flagon to refill his ceramic mug. Robert realised that he had to act quickly if he was to get his fair share of the drink, as the Scotsman wiped his mouth with the back of his hand.

"Bless or curse, nobody can decide which to do. Yes, it's all getting nasty in Rome. The word is, that they think that our very own dear departed Brother Tomasso d'Aquino,"(Robert winced at the attempt at an Italian accent) "latterly of this parish, went a bit too far. *They* being the Pope and cardinals. And *they* like to wait until their opponent is safely dead, don't they? Time's on the side of the Church. Of course, there are lots of other stories flying around, but remember that gossip about an urgent letter arriving for the Bishop, a couple of months ago? William and I joked that if any trouble was brewing, it would be just like the old days. Of course, you weren't here back in '70, probably still learning your letters!"

Robert gulped some more of his wine to reach for the flagon before his companion – no, his *friend* – could take yet more, and ended up spilling some down his tunic. He cursed under his breath; this was a waste of wine and he'd spoiled his clothes. It wasn't turning out to be a good day. He tried to smile away his irritation.

"No, I wasn't. Here, that is. But it didn't really have any effect, did it? Just banned a few of the books which mad old whatsisname, Siger, had cribbed from the Muslims. Fair enough, in my opinion. Just like the other one, back last century. So what's different with this one, or is it just a storm in a cup?"

Andrew frowned, reaching for the flagon.

"We're going to need more wine. My turn to buy. This time, the Pope's involved, like I said. And 'the other one' wasn't that long ago, only sixty-odd years, not last century."

He pushed himself heavily to his feet and turned back towards the innkeeper's table, leaving Robert alone with his thoughts for a while.

He really couldn't see what the problem was. Each time a bishop banned some books, a few students got frightened off to these new, disreputable, places like the so-called "University" of Toulouse, but soon enough things were back to business as normal. Grosseteste himself had studied in Paris when Aristotle was on the banned list, and he had gone on to be one of England's greatest philosophers! Or think of that celebrated Oxford man, Friar Bacon of Folly Bridge – he had taught a course in Paris for years, on The Philosopher and The Commentator. The young Englishman became aware that Andrew was refilling his cup, and smiled up.

"Fat lot of good it did the bishop last few times, look at Bacon!"

The Scot scowled.

"Bad authority to quote, Rob. Bacon's been arrested – letter from Oxford, just to add to the chaos. If the Pope's really behind this, it could turn into a blanket ban across all the universities. It's being enforced by excommunication. And it's your lot that's to blame."

He muttered something in an impenetrable accent. Robert let the muttering slide – he got the general idea – but couldn't ignore the slight on his profession.

"My lot? You mean the theologians? Just because you've never wanted to move to one of the professional subjects... I don't want to be stuck teaching grammar and geometry to youngsters all my life!"

Andrew smiled, leaning back against the wall comfortably.

"No, Rob, you know I'm happy in the Faculty of Arts. You're a young man and in a hurry to set the world ablaze. I've no quarrel with Theology, Medicine and Law, you know that. Between them, they leave plenty of leftovers for us poor Masters of Arts to pick over. Case in point, how the world actually works. I take it you don't despise Bacon for sticking to that?"

It was a recurring argument between them.

"But even if you don't, somebody out there obviously does. That's the other thing that's different about this time round, my friend. It's not just a few books that have been Condemned. It's a whole range of ideas that you theologians think we can't handle over on the Arts side of the fence. Including the old ones, yes, Aristotle and his Muslim disciple Averroës, your famous *Commentator*, but also a lot of modern thinking – that's new, and different, and dangerous. Some of the Condemnation refers directly to ideas from our Italian friend – did you know that they're starting to call his work heretical? Don't tell me you're going to get on your high horse about Christian teaching and pagan philosophy, Rob, you know how we can't do without Aristotle and The Commentator when we get on to natural philosophy. We're not allowed to say any more that there can't be a vacuum, that you can't move the whole universe, that the universe couldn't possibly be eternal... all these basics, that underpin everything else! If you can move the whole universe, then what follows? St Augustine must be weeping in heaven."

"Well, you don't actually *need* to do natural philosophy, only astronomy's on the syllabus, and..."

Andrew spluttered out his mouthful of wine.

At least, thought Robert, *it didn't go on my tunic.*

"That's hardly the point, and you know it. We've got to press on with new stuff as well, and over in the Arts Faculty we don't just limit ourselves to teaching. Yes, if any of your beloved Church Fathers had turned their hands to the everyday behaviour of objects around them, our life would be much easier. As it is, we've got a Pagan Greek and a devout Muslim as pretty much our only authorities, apart from a couple of obscure Jews. Even the astronomy that we have to teach is all from pagan sources. And what did the Church Fathers do instead? Make cheap cracks about how pagan learning has nothing to do with salvation, so that a thousand years later your colleagues can complain to the Pope. *What does Athens have to do with Jerusalem?* Sometimes I wonder what either of them have to do with Rome."

Rob felt that the argument could go at least three ways, and none of them would be good.

"Hey, Andy, calm down. I've got nothing against modern learning, you have to find the truth where you can, even in unlikely places. But that's not the same as flatly denying anything Biblical, and I don't see how you can have your cake and eat it too."

That wasn't quite what Robert had meant. He had a sinking feeling that the conversation was going to get veer back to Tomasso.

"That's just the thing. You can! Thomas showed that you can! You know they're already calling his supporters *Thomists*. Isn't that an awful name for such a great idea?"

Andrew's eyes were gleaming. He leaned forward, lowering his voice conspiratorially. Rob reflected that nobody in the pub would understand them or even care. What a life they lived!

"So, my fine Kentish friend, you can do it all. You can be a good Christian – even a Dominican! – and use ideas that have come from outside the Church. You just have to be damnably clever and follow the logic wherever it takes you. It works! Although maybe *damnably* isn't the best word, given the circumstances."

Robert frowned into his cup. Nothing new, nothing new under the sun. Tomasso had been a good man with a witty manner, popular with students and teachers alike, but a whiff of heresy had always hung around his immense, precise, treatises. Perhaps this Condemnation was actually the right thing to do, had nobody considered that possibility? How could he put that across without offending his... colleague?

"Even your Thomas was confused about astronomy, though. We've known that The Philosopher can't be the last word on the subject for, what, eleven hundred years. So..."

"...so, we use a different bunch of pagans instead. You're not listening, Rob. Here, have some more wine."

He leaned over the table to refill the cup, but Robert had stopped drinking some time ago. Shrugging inwardly, he filled his own instead.

"You know we need authorities to guide us, those of us who aren't intellectual giants, anyway. Can we be too picky about where we find them? Or do you think that all books should be banned, or burned, apart from the Bible?"

Robert felt tired. It had been a long few days, and the wine was making him even sleepier. He had no particular desire to defend himself now; this was just a chat between two scholars, not a formal academic debate. Surely Andy knew he couldn't believe that? It was so tempting, so easy, to let Mother Church do the thinking for you. But then what was the point of his further study, why be a theologian rather than leaving the University world to be a simple parish priest? He closed his eyes as he imagined an alternative life, living in a comfortable house, saying the canonical services, being a person of some influence... maybe not so bad, after all.

He grinned suddenly.

"So what's anyone doing about it? Or do we all just weather the storm?"

Maybe it was time to order some food, surely that would revive him. Until then, maybe a little nap wouldn't be such a bad idea. It was lovely and warm here, and so grey and rainy outside.

Andy drained his glass. Up-and-coming scholar though this Englishman was reported to be, he wasn't very bright, was he? Maybe he should give him the benefit of the doubt, but his encounters with academics from the other faculties usually left him vaguely disappointed. The best philosophers, he reflected, were universal men, with knowledge of everything, and they certainly didn't spurn the study of nature. Look at the Ancients – Aristotle himself the prime example! And his great commentator, Averroës, had to know nearly as much. In the modern age, Bacon, Aquinas, Albertus the German – all generalists. That's why he'd always stay in the Arts Faculty, teaching the seven liberal arts. Seven, not just one, and they were only a start.

Except, how could he continue if the Church really was serious about stamping out all traces of Aristotelianism, of Averroism, of (he chuckled) Thomism?

Andrew placed his cup down carefully, and stretched. Young Robert had nodded off, his head lolling forward, his fashionably long, dark, hair falling in front of his face like a curtain. The kindest thing would be to let him sleep.

He's a lost lad, a long way from home, and he's got very little experience of Church politics, let alone when they intersect University politics. I'll sit here for a while with him, finish the wine, make sure nobody steals his purse. There's nothing I particularly have to do today, and it's good to get out of that Hall. I'm no Englishman, but there are fair few Scots here, and better the devil you know...

He swilled his wine around thoughtfully, only partly aware of the loud party of youngsters who were telling loud, crude, jokes at the next table to much hilarity.

Would it be easier to be like them, to go through life thoughtlessly, not to care so much about everything?

The lad asked a good question, though. What's to be done, or will we just sit here complaining? I need to speak to Siger about it, and to Boethius. Unless they decide to call the Church's bluff, and get out of town this time like they threatened to in '70. It's been obvious that Bishop Tempier has been after them for years. What can they do, though, what can any of us do if we're forbidden from teaching that this stuff is true?

There didn't seem to be any wine left, and it was getting dark outside..

Probably better to get back and get some bread from the baker's on the way, rather than paying the inflated prices here. 'Forbidden from teaching that this stuff is true.' Not forbidden from teaching it as a possible theory – could that be it? Is the loophole really so simple, so obvious? I'll have to see a copy of the original declaration. If it's Tempier's clerks who drafted it, rather than the Papal lawyers, there could be holes large enough to pass through with all the Company of Saints, singing Alleluia.

Andrew stood, pulling his much-patched cloak around him, ideas swirling in his head.

It goes further! We can discuss other possibilities – what a fantastic exercise for the students, and for us! 'What if we could move the heavens, what would follow?' 'What if there were other worlds, what would follow?' 'What if there were a vacuum, what would follow?' The possibilities are endless. I wonder how far you could go, purely suggesting hypotheses, before the Bishop starts condemning speculation? Surely, that's a step too far... how could you ban discussion of what might be, without tying yourself up in legal knots?

He leaned over the table, shaking Robert gently by the shoulder.

"Come on, my tired young friend. Let's get you home. Don't worry. Everything is going to be fine. Here, let me lead the way."

The pair stumbled into the endless March drizzle, and passed quietly through the darkening streets.

Notes on Chapter Six

The Condemnation of Paris

Fictional characters:
Robert of Canterbury, Andrew the Scot, William

The great historian of science, Pierre Duhem, called this event "the birth certificate of modern science." As Andrew noticed in the chapter, it gave theoreticians *carte blanche* to speculate about all sorts of things previously considered impossible.

Mediaeval universities were divided into faculties. Undergraduates had to belong to the Faculty of Arts, to study the seven essential subjects. The preliminary three were Latin, logic and rhetoric – the *trivium,* root of our word *trivial.* The more advanced four, the *quadrivium,* were arithmetic, music, geometry and astronomy. Graduates could continue in the Arts Faculty (as teachers and researchers) or move to the Faculties of Law, Medicine or Theology. Most members of those faculties also taught Arts courses as a source of income.

There was a lot of mobility between universities, both for students and teachers, since the Arts curriculum was identical in all of them (and taught in Latin). Students and academics would often share lodging houses or halls with others from the same country. At the time, these halls were known as "nations" (from *natio,* to be born) – the first use of that word.

Thomas Aquinas drew on Aristotle ('the Philosopher') and the Islamic philosopher Ibn-Rushd (Latinised as *Averroës,* known as 'the Commentator') to produce a systematic synthesis of the best of Greek, Muslim and Christian thought. This was initially considered heretical, but was rapidly adopted as the Catholic Church's official philosophy.

7

Disorder

AD 1355

In which the age-old rivalry between Town and Gown turns violent, and Richard "The Calculator" Swineshead, one of the first men to use mathematical methods in the study of motion, ruminates on the usefulness (or otherwise) of astronomy, physics and philosophy in a violent and broken world.

February 11th

Richard gasped as he was flung to the ground, his long academic robes fluttering around him as he fell. *What the devil?* Muddy water from the roadside ditch was seeping into his long sleeve. There were sounds of a scuffle around him – was that the ring of swords clashing? He shook his head, and started to clamber to his feet. He was dimly aware of footsteps running away, then felt a hand grasping his arm to help him up.

"Are you alright? How are you, sir? Are you hurt?"

He peered at his rescuer as he regained his feet; he seemed to be an undergraduate from the age and the cut of his gown, a tall, sturdy, dark-haired teenager, with a strong northern accent. But since when did undergraduates wear swords on the High Street? Behind his shoulder, his blurry eyes could make out two more young men, standing protectively with swords actually drawn. One of them gave a start.

"By all the stars and saints, it's the Calculator himself! You shouldn't be out here, sir. Come on, let's get you back to your college. Merton, isn't it?"

Richard nodded.

"What –"

His voice caught in his throat, and he coughed to clear it, shaking his head again. What sort of nightmare was this?

"What is happening?"

The dark-haired one answered for the group.

"It's a full-scale riot, sir. Nobody's quite sure how it started yet but it's serious. There's been killing on both sides."

"Sides? Sides?"

"Town and Gown, sir. It's not safe for a distinguished scholar like yourself to be out on the streets. Didn't you hear the University Church sounding the alarm?"

Richard slowly shook his head. Although, come to think of it, he had wondered whether the services had been moved to new times, or if there was a special celebration of St Scholastica in progress, so at some level he had been aware of it.

"I was visiting a friend, at Balliol."

"Take my arm, sir, and we'll get you back. I'd advise you Merton fellows to keep yourselves locked up until this all blows over. You don't have youngsters like us to fight for you."

That was true enough, thought Richard, although there had been some talk of admitting undergraduates to his college. It would be a good thing for the place to get some fresh young blood, but there was so much opposition to the proposal that he doubted any change would happen in his lifetime.

"Thank you, young man. Are you from a college or a hall?"

"John of Carlisle, of The Queen's College, sir, at your service. We were just... sweeping the chaff from our threshold, you might say."

His friends smiled grimly.

"Thank you, young man. I'll make sure that your Provost hears of your good deeds. Now, where's my stick?"

Even as he spoke, one of the kind young men was handing it to him.

The little party made its way down past Mary's Hall to Merton Street. The streets were very quiet. What was the world coming to, when even Oxford wasn't safe? Richard was so relieved to see the college gates that he almost forgot the discomfort of his soggy sleeves.

At dinner, the excited fellows could talk of nothing else. Richard had found himself elevated to the status of a demi-god for having, inadvertently, ventured onto the streets which his colleagues seemed to think were running with blood.

"You say you were actually assaulted by these ruffians, Swineshead?"

He smiled modestly.

"Well, I had a little tangle. There was even some swordplay!"

The Warden looked serious as he counted off the fellows.

"I've heard that the townsfolk are sending messages to all the local villages. The Chancellor has sent for the Duke's men and the King's men. It could become even more uncivilised. We have a sacred duty to protect the college, both its material and its mortal elements. I am forbidding any of the fellowship to leave Merton tomorrow. The gates will be locked and only opened at my express permission."

There was a murmur of assent. Richard frowned.

"And the chapel door?"

The splendid, half-completed college chapel doubled as a parish church, and had an entrance from the street. Could parishioners be denied access to their place of worship?

"It will also be locked. I shall explain matters personally to the Bishop, as soon as I can."

Richard Swineshead made his way, alone, to the chapel. A snoozing verger jumped to his feet as the door creaked open, with a look of wild panic – did he think that the barbarians were invading his sacred domain? He drew a deep breath to calm himself, nodding to the academic and scurrying to fetch his materials for lighting at least some candles and lamps. He was well used to the old man's solitary evening visits. It reassured him that the man that everyone said had the cleverest mind in the land seemed his normal, unflappable, self. If The Calculator wasn't worried, then the stars could continue in their courses and the comfortable routine could be maintained.

The verger struck a spark with his flint. He shielded the reed taper with his hand against the draft, moving deftly through the dark of the chapel, *his* chapel, to Swineshead's habitual stall, lighting three tall candles. He bowed deeply as they crossed paths, one returning to his comfortable, cushioned chair and his dreams, the other following his habitual path to the one place on Earth where he could free himself from distractions, stick tapping on the cold stone floor.

Or was it the only place? Richard mused, as he had done so often before, on the relative merits of chapel and library. The latter always seemed so full, full of earnest men in a hurry to make themselves a reputation. His reputation, such as it was, had already been made and could now only be changed by the ravages of time. They called him 'Calculator', sometimes to his face, and it was a noble title. He smiled. Was it wrong to be proud of one's achievements and reputation?

He had reached his stall, and sat slowly in the intersecting circles of light, his trifold shadow flickering along the walls.

Placing his walking stick carefully on the floor, he bent forward, his arms resting on the back of the pew in front of him.

What a magnificent place this was! The Calculator wondered how many more generations it would take until it was finished. Not until after the college numbers had grown substantially; the fellows rattled around like dried peas in a child's toy, and for all his concerns the townsfolk didn't come there to pray as often as they could, or should.

He didn't need the candles to see in his mind the long pews of the Quire, facing each other across a noble aisle. He had seen them often enough in all lighting conditions, as the sun struck from all angles. He had wondered of the propriety of installing a sundial on the floor of the chapel itself, one of the pillars acting as a gnomon, but the light was too diffuse, too imprecise.

What could they do to make the townsfolk get along with them better? They saw academic pursuits as the fond delusions of men cursed with the need to question, to investigate – yes, and to calculate. They saw the university dons as customers in their stores and at the market, and as otherworldly marks for their swindles, no doubt. Two sides, town and gown.

Nobody was born to a gown, though. Even Bradwardine, may his soul rest in peace, was from farming stock, and many of the best undergraduates had risen through parish schools, for all the sons of the nobility surrounding them. He stopped for a bittersweet memory, adorning his mental image of the chapel with the finery and pomp of the memorial service for his old friend and colleague.

The actual funeral had been in Canterbury – where else? – but his college had remembered him fondly, and given an extraordinary feast in his honour that had made the Bursar sweat for days over the cost. Bradwardine had been a true nobleman of the soul, regardless of his birth, and the greatest of them all. Perhaps one day he would be made into a saint, just like Thomas

Aquinas. Scholar, Professor, Chancellor, Archbishop – his name and reputation would surely persist as long as scholarship itself survived. The question was, would it survive? Did it deserve survival, or were those who would have happily killed Richard and robbed his college's treasury correct?

"I will sing praise to the Lord with my spirit, and also with my understanding," wrote St Paul to the Corinthians, and they were singing so loudly here. If the Bishops and Archbishops were leading the choir of the spirit, did the Doctors and Professors direct the music of the intellect? Would the townsfolk know, or care, if that were the case?

Richard considered, once again, how he could justify what he had done with the last thirty years of his life. Surely, everyone agreed that we could and should understand the movements of the heavens. For the learned, there were books by Campanus. For the less learned, there was John of Sacrobosco's little textbook that the teachers somehow crammed into the heads of their students by sheer repetition. But even the unlearned appreciated the need for almanacs, for navigation tables, for their horoscopes – where would the medical doctor, the apothecary, the barber-surgeon be without tracing the influence of the planets on their patients?

Yes, the perfect unchanging Heavens had their uses on Earth, and everybody knew that, even if they were content to leave the calculating – that word again – to others. Even if Campanus had been a little obsessive, and had calculated every last motion, even ones with no practical use. If you told the butcher, the ale-house keeper, the thresher, that an Italian of the last century had studied the movements of the planets in more detail than anyone before him, you would see him smile and shrug, and he would say, "Well done!" Studying the Heavens was a universal good. Did he exaggerate? Would anyone object, or say that the time spent in this work has been wasted? No, no, astronomy was safe from the mob, at least now in this Christian era.

He lowered his eyes, his head resting comfortably on his arms. The problem was the study of the changing, mortal world below the Lunar Sphere. Everything was so unpredictable, people said. Yes, a mathematician could speak to an astronomer, and between them they could examine everything essential about the aether, but the everyday elements of earth, water, air and fire – surely they were too erratic, too chaotic, to analyse? Each particle of each element tried to reach its natural place, but the world was so full of particles that they collided with each other all the time, and their paths were no longer simple. Was it a fool's errand for the mathematician to speak to the philosopher of sublunar nature, to the physicist? Surely, Richard hoped, it couldn't be.

Look what he and the other physical mathematicians had achieved! The Philosopher himself could only talk about earthly motions as being 'slower' or 'faster' and The Commentator, indeed all of his commentators, had followed him. But in England, in Oxford, in Merton, a few madmen had dared to hold a measuring-rod against nature, and had started to claw out an understanding, however imperfect, of motion! Earthly motion, not Heavenly! All of Bradwardine's ideas, of Heytsbury's, of Dumbleton's, yes, and his own ideas too – they were shining a light where the human intellect had never previously ventured. Let the outside world treat their talk of 'final velocity', 'average velocity', 'acceleration', 'quantity of motion' as so many incomprehensible magic words – that's just what Ptolemy's peers must have said about 'eccentrics', 'epicycles', 'equant points'. The light shining in Merton was too precious to be snuffed out, and yet too delicate to be left to burn unguarded. Richard and his colleagues were the guardians, the gatekeepers to a new world of calculation.

Let the barbarians burn the library, the ideas are what's important.

His eyes were closing of their own accord. He was so very tired after such an unusual day. Was it his physical danger that led to this triple exhaustion of mind, body and soul, or the sense of mortal peril connected with these thoughts? Richard was starting to sound proud, and suspected that he had a haughty spirit. Surely he should rather consider how his work displayed love of God and of his fellow man, rather than basking in smug self-importance. The Lord had not said, "Blessed are the physical mathematicians," at least not unless the Latin of the Vulgate Bible was even more corrupt than some of these heretics claimed.

How could it be wrong to advance mankind's knowledge of the Creation? Studying the world in which God had been pleased to place him was certainly giving Him due praise. He wished that he could sing in tune, but perhaps his variations of the Music of the Spheres were just as pleasing to the Creator as the new fashion in Church music, this *polyphony*. Certainly, his pride could not be sinful, since nothing he did was a human invention, a creation of his own imperfect powers. He and the others were simply revealing more of God's purpose, and if He had wished it to remain hidden as some allege, then why had He given humans such inquisitive and powerful minds? It was because he loved his God and his fellow man, and because he wished to help the latter to learn more about the former, that he did what he did. Physical mathematics had a long way to go – many centuries passed between Eudoxus and Ptolemy – but if nobody started along a path, nobody would ever find whether it led anywhere interesting.

Let that be my defence to the townsfolk if they enquire why I spend my life like this. And let Oxford endure beyond these passing turmoils, O Lord God of scholars. I entrust these my thoughts and the secrets of my heart to Saints Katharine and Scholastica. Amen.

August 18th

John of Carlisle raised his tankard high.

"Here's to the memory of Edward of Hexham, the best friend I ever had. Damnation to his murderer!"

The other dozen undergraduates staggered to their feet, to a ragged chorus of "Edward!" More than half of the company were more than half drunk. It had seemed only right to have another wake for their friend, now that justice had finally been handed down by the King.

The college servants hurried around, refilling mugs and tankards. Ever since the riots, the students had turned inwards, to the colleges and halls. After all, the whole affair had been started by a drunken altercation in a city centre alehouse, and feelings were still running high.

John drained his ale, slamming the empty tankard down on the table as he sat heavily on the bench, peering blurrily at the other men, drawn from all the colleges and none.

All fellow students, fellow seekers after the truth. At least we know that His Majesty supports us. Those damned dogs from the Town will have to humiliate themselves every year for half a millennium, begging forgiveness and paying a fine. But that won't bring Edward back, nor any of the other sixty who were killed, nor will it heal the fear I now see in the eyes of the old doctors and professors, the fear that their lives hang by a thread. Two thousand ruffians from local towns joined in, seeing nothing but the chance for plunder. Colleges were broken into, books and treasure seized, elderly fellows cut down where they stood to defend their store of knowledge against the barbarians. And for what? Because some high-spirited young men thought that their drink had been watered down? No, that might have been the spark, but both sides have been heaping kindling up against the logs since this University began. Why do we need them, anyway? I've half a mind to quit here and move to Cambridge, it's probably worth living in a swamp to be in a place where the University is in complete charge.

He staggered to his feet again. He had to relieve himself. He waved away the mugs being thrust in his direction as he stumbled towards the privy closet.

They killed my best friend. They're no better than dogs, and deserve to be put down. It's a damned shame I only got to kill one of them, I can still see his smug, ignorant face as he fell.

Someone was holding his elbow as he urinated, steadying him. He smiled his thanks, half-recognising one of Edward's friends from Mary's Hall. He thought that they studied astronomy together with someone pretty senior, not really an astronomer himself.

"John, Edward and I... we went for lessons with The Calculator, old Swineshead from Merton. I heard this morning... he died in his sleep yesterday. He was a good man, John, and this whole business just finished him. He couldn't hold it together, retreated to his room to prayers... and now... They're the only two people I really trusted, John. What do I do now?"

John pulled up his hose.

"We all do what we have to, James."

He had no idea whether that was the right name, but it seemed a good thing to call him.

"We'll avenge Edward, and Swinesfoot, and all the others. We'll show the world why we're here, and that they can't do without us. We'll use ink, not blood, to make our names. They'll be talking about us for the five hundred years of the city's penance, and beyond."

"Are you alright?"

James, if that was his name, was looking at him curiously. John became aware that he was waving his hands around rather wildly. It was probably time to get to his chambers. He smiled broadly.

"Yes, I think I am."

He waved expansively to all his friends, his brothers, his peers. "But I've got some thinking to do. Goodnight, everyone."

Sleep. Sleep would be so good right now. And then... work. Work as I've never really worked before. Otherwise what does it mean? I bet The Calculator never had these worries. He must have been young once, been like me, been just as drunk, though. Ah, here's my bed... goodnight, everyone. Goodnight.

Notes on Chapter Seven

Swineshead

Fictional characters:
John of Carlisle, Edward of Hexham

Swineshead was arguably the greatest of the Merton Group, several generations of academics based at that Oxford college. Merton had started to specialise in research rather than teaching, and at that time only allowed graduate members (most of whom came from Balliol, a large teaching college of the time). Many undergraduates of the time did not belong to a college and arranged lessons privately.

The Merton Group were revolutionary in attempting to apply the sorts of mathematical models normally reserved for astronomy to earthly movements, giving a huge boost to the development of physics. Unfortunately, the mathematical notation of the day was cumbersome and measurements of time and distance were largely inaccurate.

The St Scholastica's Day Riot was entirely real, and was devastating to the young university's international reputation. An engraved stone at Carfax, in the city centre, marks the place where the riot began. The annual penance and fine, to be paid by the city to the university for five hundred years, ended just a little early when a Victorian mayor decided that it was rather silly. Swineshead's date of death is uncertain (some time in the 1350s or 60s), and there is no record of his direct involvement in the events, so their connection is a dramatic invention.

8

Authority

AD 1372

In which the French bishop and all-around genius, Nicolas Oresme, theologian, economist, translator and scientist, explains fourteenth-century state-of-the-art science to his patron King Charles V, showing a healthy and atypical contempt for using astrology to predict the future.

Royal Chaplaincy
Palais du Louvre
Paris
1st July, AD 1372

Your Majesty,

Many congratulations indeed on the most richly deserved victory! Those who know of such matters – and everybody in the Court is suddenly an armchair general or admiral or both – assure me that with the continued naval co-operation of the Castillians, and the decisive leadership of Constable du Guesclin, you will be able to turn the Battle of La Rochelle into the start of a glorious campaign to evict the English once and for all from the western part of our land.

Unfortunately, Your Majesty, I am as worried as I am pleased. As Your loyal subject, and as Your adviser on theology and philosophy, and as Your chaplain, and most of all as Your friend, I am deeply concerned – not by the victory, of course, and only incidentally by the fact that affairs of state have (rightly) interrupted our regular dinner-time conversation to the extent that, even when You are in Paris, I only ever seem to see you for Confession, with no chance to talk of wider matters. No, what concerns me is the attitude that even the highest nobles and councillors are showing. They are praising God for our sudden turn of fortune, and rightly so, but they are also declaring that our country's destiny is clear from the positions of the stars and the planets. In the past, knowing Your views and my own, such thoughts have been muttered in secret. Now they are a common topic of table-talk, although those who know me best have the common grace to look ashamed of their sudden adoption of judicial astrology.

Sire, in Your temporary absence, a decree affirming the insidious nature of such talk would be of great value to those of us who are attempting to keep Your court a rational and orderly place. If our fate is foretold in the heavens, there would seem to be little motivation for us to exercise our God-given talents and abilities. Further, we would not be inclined to praise or blame others for their successes or failures, which would undermine the entire foundation of honour within the State. Indeed, the whole notion that we are responsible for our *own* actions, the very core of the Christian faith, would have to be rejected.

I know that many Church colleagues and indeed superiors of mine support the idea of a direct and predictable influence of the planets on humans, and that the Holy Father himself in Avignon is at best ambiguous in his teaching, so I wish to outline some of the objections that I have formalised. (Imagine this to be one of our fireside evenings, wine in hand, although You are probably in a tent in a muddy field at the moment. I hope that at least You have decent wine!) The astrological disease is so widespread that I am currently composing a little book on the subject. I assure You that this is purely a side-project and will in no way interfere with the translation work which You have been pleased to command of me – indeed, Aristotle's *Politics* and *Ethics* are already complete and ready to be copied into the library, with the *Heavens* and *Economics* to follow shortly. The last of these raises interesting questions about the nature and value of money itself, but perhaps that is a topic best left for another time.

In the past, I have advised You that the study of mathematical astronomy is not suitable for Princes or Kings, and I hold firm to that view. However interested You are in the motions of the heavens, command of the details requires long years of immersion in the finer points of mathematics and natural philosophy, ideally to the exclusion of other concerns and responsibilities. I have, perhaps, ignored my own advice in pursuing these matters concurrently with the study of theology and advancement within the Church, but if I am ever preoccupied with abstruse questions to the extent where I neglect my duties (as I confess sometimes occurs), a kingdom is unlikely to fall as a consequence! So I crave Your indulgence, from afar, in bombarding You with words and topics which might prove less familiar than those aspects of Aristotle which we frequently discuss. I hope that their impact will be far greater than that of any actual physical bombardment that the English might attempt upon Your position.

Which reminds me – I have recently started applying my trick of representing physical quantities by drawings to the study of motion, and believe that I have come some way to proving the speculations of the Oxford scholars (whom I cannot see as our enemies even if our kingdoms are at war). Quantity of velocity can be represented by proportional blocks, or strips of paper, and a means of demonstrating acceleration and distance covered established by such drawings. This might prove of use in gunnery as well as in the study of natural motion, and I look forward to showing You my findings upon Your return. You see that I am not merely an idle old man sitting daydreaming at his prayer-desk, or frittering away his time sampling all of the delights of the Royal cellars!

I am wandering away from my due course, though. The fixed stars and the heavenly wanderers (for that is what *planet* signifies in the Greek) certainly move in predictable ways – which I shall describe for you in a matter of minutes. It is undoubtedly true that their positions influence the course of events here on Earth, as all parts of Creation are entirely interconnected. I might add, if they do not serve some purpose, we might wonder: why did the Lord create them and ordain their movements in the first place? What I dispute is the commonly accepted link made between the two – the notion that we can understand the influences and use the predictions of position to make precise predictions of how events will unfold. I am not, as Your Majesty knows, an enemy of such an ancient and respected art as astrology (which can certainly explain the influence of the months and seasons upon our climate), merely of its misapplication on the scale of individual men and women.

You might ask why I doubt the competence and reasoning of mortal men, at least those of our present age. Surely, we have understood the motions of all of the objects in the sky – why cannot we also use our mental powers to understand their influence upon us? The easy answer attacks the second part of this question: such influences are so weak that they merely *suggest possibilities* rather than definite certainties, and it is notoriously hard to analyse or discuss the merely probable rather than the necessarily true. We are all aware that when the medical doctor studies our horoscope, it is merely one of many factors which he uses to diagnose our ailment.

However, it is the first part of the question which provides for a more interesting discussion. Do we know with absolute certainty how to describe, explain or predict the motion of the heavenly bodies? At the risk of undermining Your faith in my astronomical brethren, I would have to answer with an understated and qualified "no."

This perhaps surprises You, as indeed it would surprise my colleagues who produce immaculately precise tables of planetary positions based on the continuous traditions of a thousand years. I do not believe that I have gone mad in my hours of prayer and parties (if You believe that I indulge in such), and will endeavour to explain to You. (Is it selfish of me to wish this war over, not only for the good and the glory of France, but also so that we can resume our delightful discussions of such things?)

There are philosophical considerations involved here, about the nature of knowledge itself. When can we be said to know something completely? Many would hold astronomy up as a perfect example of a subject in which we can know precisely the causes and effects, in which we can be confident of our learning. This is not, alas, the case. We educated men all know, from our University studies, that Creation is composed of spheres, made of aether, and somehow, within those spheres, the planets rotate around little circles called epicycles, themselves moving around the Sun, and our doubts about the complexity of this combination are somehow assuaged by the comfortable language, spoken with such authority by our teachers. Yes, I sweated through Sacrobosco's textbook as You may have done. How, though, do the spheres relate to the circles? Aristotle would have us believe one description, and Ptolemy another. Rabbi Maimonides described the compromises made by Averroës' teacher, one Avempace, but they lacked mathematical sophistication. Later,

Friar Bacon (those Oxford men again, one would almost think that other countries lacked any scientific vigour) placed the circles between the concentric spheres. In antiquity, Proclus believed that the spheres were real but the circles were mere figments of astronomers' imagination. History does not record what Ptolemy thought about the truth of his explanations. I, myself, incline to Bacon's view, that the epicycles and eccentrics must exist, and must fit between the Aristotelian concentric spheres. But the long and the short of it is: astronomers differ in their interpretation of the theories we all learned at college.

What value does my voice have, amidst those of such great authorities who made the heavens their lifetime study? They could not agree amongst each other about the chain of cause and effect, about the influences and mechanisms involved, even though they would happily sit down and calculate the positions of the planets on any given night with assured precision. Averroës (alas) generally simply states that the Aristotelian spheres are sufficient explanation by themselves, not realising that this answer raises yet more questions. In some of his writings, he goes further; in his commentary upon *On the Heavens*, he claims that he will explain how to reconcile the theories – but finding himself unequal to the challenge, he silently breaks his promise and leaves the reader in despair. And even Aquinas, normally so pedantic, glosses over this point, merely claiming that one or the other explanation must be correct. Why is he so sure? Perhaps the correct answer is one which has not yet been proposed, which must await another age. I claim no credit for this observation, which although it must have occurred to many previous thinkers, was first recorded by al-Battani in his thoroughly sensible Commentaries.

If we cannot understand the chain of causation in the heavenly movements, which have been recorded, measured and scrutinised these past few thousand years, what hope could we ever have of capturing the infinitely more subtle influences exerted by motions in one sphere on those in another?

That might seem bad enough to Your Majesty, but the uncertainty about the foundations of astronomy runs far deeper than this. Everyone agrees that the Earth is stationary and at the centre of the universe, and that the stars are fixed points on a sphere which rotates once daily (or nearly so) around us. Those who suggest a rotating Earth are laughed to scorn, and packed off to the madhouse. Why? What arguments can we give to convince our madman that the Earth is fixed?

"If the Earth were rotating, there would be a great wind."

Nonsense! Why would anyone suggest that the elements of earth and water rotate, but not those of air and fire? That is contrary to all reasoning, since we believe that in certain essential ways the four worldly elements have the same natural motions. If the Earth rotated, the atmosphere would rotate along with it. When in a closed carriage, one does not suffocate due to air being left behind, because it shares the carriage's motion.

"If the Earth were rotating, it would have to move at an extraordinary speed."

Nonsense! If the Earth is fixed, the stars would have to move a speed many millions of times greater, and nobody considers this to be implausible. (Campanus of Novara calculated this, but he was obsessed by numbers and calculated much that is of no use to anyone. I cannot recommend his work.)

"If the Earth were rotating, we would be thrown off!"

Nonsense! Since our bodies are made of earth and water, we are pulled to our natural places (i.e. towards the centre of our universe, which also happens to be the centre of the Earth) regardless of what other motions may interpose themselves. We experience this as a force pulling us downwards, i.e. towards the centre of the universe. Would hypothetical people living south of the equator simply fall off? We who live in France do not, when we venture north or south of our temperate climate, see people standing at odd angles. When standing on a moving boat, we are not left behind but are swept along with it.

So You see that there are no rational grounds for believing that the Earth is stationary, but nevertheless we all believe that it is (myself included, lest You now be worried about my sanity). Doubtless, one day some man of genius will establish the reason *why* the Earth is fixed, and be able to demonstrate it by rational argument. That day has not yet arrived – and we are left in the horrible situation of not having any firm foundations for the elaborate cathedral which we have painstakingly constructed. Again, if this is the case for the mathematical and technical branches of astronomy, how could anyone believe that so-called 'judicial' astrology, the application of astrological theory to individual destinies, is on a firm footing? You will find astrologers more dogmatic than philosophers or even theologians, insisting that the antiquity of their authorities removes any need for independent intellectual enquiry on their part. Cheap charlatans and frauds, Your Majesty, and it grieves me to see the foothold they are gaining in Your palaces. Where there is so much light (and Your library is a wonder of the modern world) it is no surprise that the shadows should be so sharply defined.

Perhaps it's a consequence of my own academic pedigree that I'm so sensitive to these differences of opinions – I'm sure that I've told You all about the battle waged between my own teacher, Jean Buridan, and his deadly rival William of Ockham (yes, another conflict between the French and the English, waged with ink rather than blood, which on the whole is probably preferable). Such an upbringing has made me question the nature and value of authority perhaps more than is wise. Your Majesty will be pleased, I hope, that my translation of Aristotle's *Heavens* contains a certain number of my own thoughts, for I fear that the unquestioning acceptance of The Philosopher's opinion is as stultifying to philosophy as the similar attitude towards the ancients of astrology is proving to morality and ambition in Your Court. Although the Holy Father would doubtless disagree, even Aquinas consistently errs in his adulation of his great teacher.

As I get older, and see more of the world, and read more, and more, and more, I am increasingly convinced that some of the very words that we use to describe nature are ill-defined, slippery, perhaps meaningless. I am certain that Aristotle has shed light on very many problems, and has helped mankind along its path as much as any mortal, but I am equally sure that if we are, as a society, to progress then we must leave him and his commentators behind us. The English have taken the first steps in this direction, with their new definitions of velocity, acceleration, motion – surely, key ideas in the description of the world. I myself have started to doubt the way in which we have traditionally defined 'time' and 'space' in terms of a body in uniform, natural, unforced motion, thinking that this is putting the cart firmly before the horse. Are we not able to establish the meaning of these words without reference to particular objects?

If the débâcle of the Condemnation of Paris in the reign of Philip III showed anything, it showed that we must consider all alternative systems. I find that there are entirely reasonable and conceivable propositions which simply cannot be expressed using the strict definitions of our forefathers. To me, it is a perfectly sensible (if odd) idea that there are other worlds in God's creation, other systems of spheres separated from ours by void and possibly supporting other beings. Each such system of spheres would, of course, have its own laws for natural motion governing whatever elements may be found therein. However, when I come to frame such suppositions in the traditional vocabulary, I find that the very definitions of space and movement prevent me. Should this mean that I abandon my hypothesis? By no means! Rather, I should abandon the constricting mesh of words which hinder, rather than aid, my philosophical conjectures.

You will excuse these my ramblings, Your Majesty. It has been a long day here, and I have dined well. The Ancients and Moderns alike agree that the world is a simple place, that God is economical in His methods. All I can say is, to this ageing priest it seems that the world is ever more complex. It is entirely possible that the numbers involved in the motions of the planets are, mathematically speaking, irrational – this would entail that the patterns of the planets against the night-time sky would never repeat themselves precisely, but spin through every possible combination until the Last Trumpet. Sometimes that is how this world of mortal men seems to me, and it is our misfortune to be living in an age which seems more irrational than most.

If nothing else, I hope that I have diverted Your thoughts from the incessant stream of logistics and orders and plans and maps and details which must assail Your Majesty hourly. I urge You to return soon to Paris out of pity for Your poor friend Nicolas, who otherwise will fill all his hours scribbling letters on overly-expensive paper.

With all my duty and affection,

Your humble subject, Nicolas d'Oresme

Notes on Chapter Eight

Oresme

Fictional characters:
None

Oresme is representative of a whole generation of Late Mediaeval thinkers who were interested in nearly every branch of learning, including astronomy – one of the consequences of the university system. The Renaissance didn't have a monopoly on Renaissance Men! Little is known of his early life, but it is likely that he came from very humble origins, working his way into university through a series of scholarships. He eventually found a loyal and generous patron in Charles V of France, ending his life as a respected bishop with a string of books to his name (original works, commentaries on classical authors, and translations).

He came within a hair's breadth of inventing the concept of the mathematical graph. His early attempts effectively created a bar chart of some property, such as the temperature measured every inch along a heated iron bar.

Oresme contributed to technical astronomy, and to cosmology. Although his ideas weren't revolutionary, he is interesting for his frequent and fluent attacks on astrology. These are unusual for the time, but fell on deaf ears. As John North put it, "Oresme's influence on astrology was hardly greater than the influence of Zeno [who 'proved' that all movement was impossible, using an arrow as an example] on men who shot arrows."

All of the thoughts expressed in the chapter can be found in his work, including the defence of the possibility of the Earth's motion (even if he didn't believe it himself).

9

Remembrance

AD 1473

In which we see the life and works of Cardinal Bessarion, the man who more than any other linked the Greek-speaking East to the Latin-speaking West, and who provided a whole generation of scholars with material to study, reflected in the conversations of those who knew him.

The Doge's Palace, Venice

"Pass me that plane, Giuseppe."

The young apprentice turned to the workbench set up in the nearly bare room, the neat rows of tools aligned with the orderly precision that was his master's hallmark, as much as his soft voice and angular features. Sometimes he thought that this mania for having everything in its rightful place was simply an affectation to produce more work for the junior members of the workshop; they were allowed precious little opportunity to do anything practical!

"The large, the medium, or the fine, master?"

He knew even as he asked that he had failed some test. If there had been any doubt, Master Sandro would have asked more precisely. Yes, the older man was straightening, turning, a look of bemusement on his face.

"Look, lad, at the wood I'm working. Why would I want anything but a fine plane for finishing the piece? What precisely do you think we're making, a pew?"

Giuseppe blushed as he handed over the right tool. Although he felt honoured to be Sandro's only assistant on this particular project, he missed the rough and tumble atmosphere of the cathedral yard, with all the other tradesmen (masters, journeymen and apprentices alike) cheerfully insulting each other's competence and parentage.

Sandro cast an experienced eye along the blade.

"At least it's sharp, and that's something. Come and watch, lad, and see what you can learn today."

No, the atmosphere here was completely different, and too intense. They had set up on site, in the group of rooms that was to hold all the cabinets and shelves. Through an adjoining wall, gusts of laughter from the Ducal Guard's barracks could be heard. Being a carpenter was a good trade, everyone said so, and he was on course to join the Guild, and maybe even a Scuola. He'd like to be part of one of those – socialising with his betters, doing charitable works, helping run the neighbourhood, eating huge feasts! If Sandro dropped the right word at the right time, he could stay here working at San Marco's for the rest of his life. If only he could focus on the delicate strokes of his master... but his stomach was rumbling, and his thought was more of the lunch he had carefully packed before dawn.

Sandro's strokes were smooth and delicate, the rough elm starting to look far more like a finished piece of... what was this going to be? Giuseppe ventured a guess.

"Is that the front of a drawer?"

His master smiled.

"Of course it is, youngster. Shelves for the books and individual drawers for the scrolls."

He looked theatrically around him, a look of disdain on his features.

"It's hardly our fault that the surroundings won't match the quality of our work. Give it a few years and this library will be so famous they'll get someone in to make a new building for it, mark my words. All the more work for us when it has to be fitted out, though if I were the treasurer I'd do it all in one go."

He placed the planed wood carefully to one side, next to the similar pieces he had been working on while Giuseppe had been sawing up the planks, and handed the plane back to his student.

"You really think so? Something special, then?"

"You know I'm not book-learned myself, but I respect those who are. It's a noble calling. And Bessarion has certainly left the City a gift and a half to remember him by! We'll be the envy of the world."

"This is all due to the Old Greek?"

Sandro laughed. "We've been working a week on fitting out an extension to the library, and you haven't wondered why?"

"I don't get paid to wonder, master."

"You don't get paid at all, lad!" This was an exchange familiar to both of them.

The Campanile bell struck two, and master and apprentice alike sat cross-legged on the sawdust-strewn floor, unpacking their lunches. Giuseppe peeled his boiled egg, trying to think what he knew about the flamboyant and popular Papal Legate to the Serene Republic, Cardinal Bessarion the Greek. He was always at the forefront of processions, ruddy face masked by an impressive beard, and had a reputation for being not only clever but also cunning, a combination essential for the highest ranks of the Church. They ate quickly, the silence between them now more companionable.

"I thought that Bessarion gave his library to San Marco's a bit after he arrived – I'm sure that there was a feast and that there was wine for all the cathedral staff, even a young nipper like me. I remember these things."

Sandro nodded, wiping his mouth free of breadcrumbs.

"Pass me that flagon, lad. Just good honest beer for us today, though! Yes, but it turned out that he had even more to give, he was just waiting to see how things turned out here. Now he's off to his new job and he's added the rest. So much stuff that they can't fit it all in the old library."

He glanced over at the sketch plans for the whole project.

"I reckon close to a thousand books and scrolls, all told."

Giuseppe whistled.

"A thousand books? I didn't even know that there *were* a thousand books in all the world! How did he get them?"

Sandro leaned back against the wall, taking a large bite from an apple. He chewed it thoroughly and swallowed before he spoke, fastidious as usual, letting the question hang in the air.

"He brought them with him when he came West, before the damned Turks got Constantinople."

He spat out a pip eloquently.

The young man nodded.

"And they're... valuable?"

"More valuable than you can possibly imagine, lad. Even though you can imagine a lot of wealth. Trouble is, they're all in Greek – it'll take years to translate them into Latin. The old boy is doing his best to help out, but now he's off to France. Big loss for the City, I think."

He jerked his head over to the bare wall where the main run of shelves would stand in a few weeks.

"I tell you, lad, these rooms are going to be a treasure-trove. Stories, history, philosophy, physics, medicine, astronomy, maths – the full works. Things that we haven't seen in Italy for hundreds of years. Folks will certainly remember old Bessarion! Are you just going to leave those eggshells on the floor? Have a bit of respect, lad. Remember, a true craftsman always has dignity and respect."

"Sorry, Master."

An artist's workshop, Venice

As he leaned in carefully to inspect his newest apprentice's work, the Campanile's bell striking the third hour of the afternoon, Gentile Bellini allowed himself a quiet smile of satisfaction. Although the task was simple – applying a dark wash to the prepared wooden panel – the youngster had done it conscientiously and systematically, achieving a completely even tone. He had stayed exactly within the space indicated by the soft lines on the gesso base.

"Yes, that's it precisely, Luca. Now, if you'll bring me my charcoal, you'll see how I mark out the areas to be reserved for the fine detail, before we apply the next background coats. Here, pull up a stool."

Bellini was always pleased when the really young apprentices – and this one could be no more than thirteen – showed diligence and learned well. Unlike many of the Guild Masters, he actively enjoyed his role as a teacher. The artistic fireworks and experimentation could be left to his younger brother: Gentile's job was to ensure the smooth running and continuity of Bellini and Sons, and that meant giving the patrons exactly what they expected. Which included the name of the firm, although with father long dead The Bellini Brothers would have made more sense.

This Luca was obviously destined for great things. He had noticed the master painter's furrowed brow as he quickly scanned the cluttered workshop, and had guessed what was missing. Along with the charcoal, he brought over the folder containing various sketched profiles and full-face portraits of the heads and shoulders of current and past patrons. Bellini nodded his satisfaction as he gently sorted through the pages, finding the distinctive outline of his friend, John Bessarion.

Even the name bought pleasant recollections.

"Call me John," the churchman had asked in his thick accent at their first meeting, after the painter's ill-fated but sincere attempt at pronouncing his real name. "It's easier for everyone, and you won't make me wince."

The ready smile had quite won him over to the new Legate's side, and nothing after that had dented his opinion of the man.

Gentile flicked his eyes between the paper and the panel as he sketched quickly, and the prominent nose and beard formed on the shape of the man at prayer. Luca clapped his hands in delight.

"Why, it's The Greek! You've caught him perfectly, sir. May I ask you a question?"

"Of course you may. Hard workers who care about their art are always welcome to ask me questions, my lad."

He was happy in his work, he thought, and the mortgage would somehow take care of itself. He was sure that the money that patrons owed him would arrive on time, and it was such a beautiful spring day. How could anyone be sad?

"Why's the panel such an odd shape?"

An excellent question. Some of the youngsters in the workshop only cared about their next meal and did what they were told without thought. Actually, some of his fellow masters were like that!

"It's to be a reliquary door, my lad. The Greek is donating a very holy relic to the City, as a mark of his thanks for our hospitality during his appointment here."

And as a mark of thanks for his particular hospitality, the food and wine shared under the stars in his courtyard, and the endless, endless questioning about the history of art in the Latin world – well, as his thanks he had been given this commission. A painting that would be worthy of the relic, and which would cement both of their names in the history of the Republic. And a fee for it that was well over the odds!

Luca gasped.

"Not the fragment of the True Cross? I thought he'd only loaned that!"

The middle-aged painter turned to his young student and smiled broadly.

"He had loaned it. But when he saw what good care the Scuola della Carità took of it, he left it as a present, along with some books."

This delighted Luca.

"Oh, I love the procession of the True Cross, and now we can have it every year."

He actually clapped his hands with joy, much to Gentile's delight. It was always reassuring to give the lie to the perpetual talk of how kids today were all disrespectful louts with no religion or learning.

"Is it true that he brought it with him all the way from Jerusalem itself?"

Gentile smiled.

"Not quite, my lad, just from Constantinople. You know, before he came to Italy, he was their – well, I suppose it's something like a Cardinal Archbishop. The things he's told me about the East, the Greek world! Although I suppose we should call it the Turkish world, now. I'd love to see it someday – imagine the colours, the shapes, how different everything would be from Venice for a painter! He paints such a picture with his words, Luca, he's a brilliant speaker. I appreciate craftsmanship wherever I find it. The Republic will miss him greatly, but will remember him fondly."

Luca nodded wisely.

"I'll certainly remember his name, master, and tell my children that my master knew him, every time we see the True Cross in processions. You're right, his name and memory are safe with us!"

Poor old John Bessarion. He had come to Italy with such high hopes, so many plans and schemes – for the reunion of the Catholic and Orthodox Churches, for the uniting of Latin and Greek scholarship, for the blending of Western and Eastern philosophy and art. Then when events overtook him, for reconquest of his beloved home, and the re-establishment of the Byzantine Empire. Was this how he would be remembered, as the founder of an annual procession and feast? Still, that was more than most people manage.

Gentile smiled at the enthusiastic boy.

"Yes, by all means remember him, Luca. Promise me this, though – you'll remember what he did, and tried to do, with his whole life, not just his years in Venice? Here, pull your stool closer and let me tell you some stories. We've done enough work for now."

A publisher's office, Nuremberg

"Otto, have a look at this. You might be interested!"

Johannes de Monte Regio placed the long letter down on his apprentice's desk. It had arrived last night, and been a fittingly cheerful conclusion to a good day's work. The new printing-press had arrived, and first impressions were that it was at least as good as the existing machine. Just as well – they had that new history book to get through before the author knocked their door down, and they had now bought almost enough numerical type to run off some astronomical tables.

It had been an excellent decision to take on an apprentice. Young Otto was showing promise as a mathematical astronomer and had leaped at the chance to work alongside the famous *Regiomontanus,* a far better preparation for his career than the conventional route of university (which was beyond his family's means, in any event). The Nuremberg publishing scene was booming, too, and Otto was showing a keen eye for finances and marketing (although it would never do for him to meet clients or customers directly; as yet he was distinctly lacking in social skills). Yes, the business was in safe hands with him. Johannes had taken on the work from Peurbach, his mentor, and was preparing his successor in turn. And so the world had turned for the last few thousand years, cycles of growth and teaching and learning.

Johannes sat in the comfortable chair, in the corner of their shared office above the press, that he used when he had no particular work to do but was amenable to conversation. He made sure that his staff, his colleagues, his clients, his apprentice had access to him for at least an hour a day, and often found this the most productive time. He turned his head to gaze out of the window at the packed street (market day caused such chaos in the streets, with over-laden carts bashing against the corners of buildings despite the protective stone bollards; something should be done), letting his mind wander while the teenager picked his way through the cramped handwriting and slightly archaic Latin.

Otto finally raised his head. He was meticulous, and liked to finish reading before asking questions, a trait strongly encouraged by his master.

"So Bessarion has donated all of his manuscripts and printed books to Venice. That's a noble act, from a noble man. It's a shame that I never met him. Do you think I ever will, sir?"

"It's possible, but unlikely. As you see, he's off to France on Papal business. You know, he was nearly made Pope himself, but the Italian cardinals put their differences aside to block the appointment of a Greek. Sad, so sad, when he spends all his time trying to patch up the differences between people. I owe him much, Otto. Come to think of it, I should pass some of that on to you.

How would you feel about learning Greek?"

Otto grinned. "It would only be of any use if we had some Greek texts to study, sir. I've always wanted to see Venice."

"And one day you will. Oh, Bessarion's library is an extraordinary collection. Some of the scrolls are over five hundred years old, the best copies we have of Homer, Plato, Aristotle, Ptolemy, Theon, Proclus and all the rest. I thank God every time I think of them – had he stayed in Constantinople and been killed when the city fell, they would be lost to us now. Instead, he gave his time, his friendship to Master Peurbach, to me, to artists and philosophers and clergymen, to teach *us* to read them, to appreciate them. We've been stagnating here in the West, Otto, and it amuses my sense of the ridiculous that the fresh thinking needed to reinvigorate us comes from five-hundred-year-old copies of two-thousand-year-old ideas. Yes, I think it's time for you to learn Greek. I'll dig out my notebooks tonight, teach you in the same way that he taught me."

Otto nodded his assent, quietly smiling. This was exactly the sort of thing he had been hoping for. His family and friends all thought that he had been crazy to accept the eccentric astronomer-cum-publisher's offer, but he knew that his natural talent was for mathematics, and how else was he going to exercise his God-given gift? Maria would laugh on the other side of her face when she saw how lucrative the book business was going to be once it got going, it was definitely the trade of the future. She might even deign to talk to him, transfer her attention from the merchants' boys and journeymen painters she spent her time with… and maybe they could become closer friends… he pulled his mind back with an effort to the here and now, the jovial features of his employer, teacher, friend whose ideas always ran away with him.

"…and then, of course, you can attempt the *Almagest* itself directly. The original is like a stream of pure, clear, mountain water compared to the muddy and stagnant versions we have from the Arabic, yes, even compared to my own summary. I'm not sure that your trigonometry is good enough yet, we'll have to work on that alongside the Greek. You did finish reading that chapter from *On Triangles*?"

"Yes, sir. In fact, I have a few questions about the tangent function, if you have time?"

Only Regiomontanus could have thought *On Triangles of Every Form* could be a good title for a book, thought Otto. When he was in charge there, things would change a bit. The company would need to reach out to the people if it was to be successful, and to print books that his family – or Maria – would want to buy and read.

"Of course, my lad, of course! Why don't you pop down to the buttery and draw us each off a nice flagon of beer? Thirsty work, this thinking."

The apprentice pushed his chair back and slipped out of the door. Johannes leaned back in the chair, closing his eyes.

The torch handed to me is safe for another generation. Wheels are turning within wheels, on Earth as they do in Heaven. Amen.

Notes on Chapter Nine

Bessarion

Fictional characters:
Giuseppe, Sandro, Luca, Otto, Maria

Cardinal Bessarion is one of those fascinating figures who only seem to exist in the footnotes of other people's stories. His biography was as described in the chapter. He first visited Italy as part of a delegation from the Greek Orthodox Church, aiming to reconcile that denomination's differences with Roman Catholicism. The mission succeeded beyond all expectations, and the two Churches were (briefly) re-united.

In the dying Byzantine Empire, Bessarion had been known as "the most Roman of the Greeks." After accepting a position as a Roman Catholic Cardinal, he was (inevitably) known as "the most Greek of the Romans." The Empire finally fell to the Turks in 1453 and a stream of scholars made their way westwards. Bessarion was already there, and became the centre of a group of intellectuals from both cultures. He travelled extensively, attempting to raise support for a formal crusade to recapture Constantinople.

Bessarion included in his circle painters, writers and scientists and was passionate about teaching them Greek so that they could work from original documents rather than translations of translations.

Gentile Bellini's reliquary-door is now in the National Gallery in London. Bessarion's manuscript collection was eventually re-housed in the fine, purpose-built library next to the Campanile and opposite the Doge's Palace in Venice.

10

Death

AD 1476

In which we visit Rome to puzzle our way through a murder mystery, and discover how the eccentric, precocious, German astronomer, mathematician and publisher Regiomontanus, met at the end of our last chapter, became enmeshed in the murky world of Renaissance Italian literary rivalries and feuds.

July 6th

Rome: the Palazzo di Modena, an exclusive hotel

Michael Benevento pushed a hand through his immaculate, short, dark hair, feeling the sweaty stickiness. It was another damnably hot July day. He smiled tightly as he looked down at the thin, pale middle-aged man with the round face, sprawled on the bed that nearly filled the well-appointed room. The man didn't smile back, but then he *was* dead.

"So what do we have here?"

The palazzo's steward was busy opening the shutters, trying in vain to get some flow of air to clear the noxious odours.

"An astronomer, sir, summoned to the Papal Council. It must be a busy time of year for you."

Michael sighed.

"Three dead this month, and it's only the sixth. What can you do?"

The steward nodded sympathetically.

"It happens every time the Tiber floods, I'm sure it carries evil influences. Can we move the body?"

"Certainly, go ahead."

He glanced down at the scrap of parchment in his right hand.

"I'd better see this secretary of his. Do you have a room that's a bit less... pungent than this one I could use?"

"Certainly, sir, I'll show you there myself, and send you young Otto, the deceased's assistant. His Italian's not very good. In fact, it's awful."

Of course it was. Everything about his job was awful. Tying up loose ends for the Papal Chamberlain's Office paid the bills, but only barely. Some of his colleagues enjoyed this life, but promotion to a desk job, even a scribal one, couldn't come quickly enough. Or perhaps one day he'd meet a cardinal, a nobleman, impressed by his perfect clothing and manner who would hire him for his personal staff.

"No problem, I'll get by. Thanks for letting us know so quickly, His Holiness appreciates your duty and devotion."

The words were a formula that even he couldn't pretend to infuse with energy. He fished an equally formulaic silver piece from his purse and tossed it to the bowing steward. Now to get out of here.

"And chilled water for us both."

There were at least some perks of associating with the rich and powerful, even as a functionary.

As Michael settled into the comfortable chair in the neighbouring chamber, he read through the rest of the parchment, the brief that the steward had been given. If only these scholars didn't have so many names... almost as bad as the Church hierarchy, with their endless titles. It seemed that Johann Müller, a.k.a. Johannes Molitoris/Germanus/Francus de Monte Regio, a.k.a. Regiomontanus had arrived in Rome from Nuremburg just under a month ago. Michael decided to think of him simply as "JR" for convenience. He was being lodged at Papal expense, was about forty years old, was a mathematical genius, and liked pickled fish and strong ale. Not together, Michael hoped, but you could never tell with these Northerners.

JR travelled with only one servant, a general factotum-cum-apprentice, who would have to be this lanky young man now hovering nervously on the doorstep. About his own age, early twenties, and scruffily dressed in a faded jerkin and leggings that looked more patching than cloth. Long, stringy hair. By the Saints, where did these people hide in daylight?

"Signor Benevento? I'm Otto."

Michael smiled easily, waving to one of the other chairs.

"Just Otto? No long Latinised name?"

Otto shook his head as he sat.

"Not yet... not famous enough!"

At least his Italian was tolerable, whatever the steward had said. It was a little guttural, but in his line of work Michael had grown accustomed to dealing with all sorts of foreign accents.

"That's a relief, then! I shouldn't keep you long, I'm really just here to do some paperwork, arrange for news to be sent home to next of kin, see if there's a will, all that sort of thing."

"Of course. I'll do anything I can to help."

Otto poured out a goblet of water, fetched up from a cold, damp cellar, beads of dew forming on the cold silver. He looked nervous, but that was entirely natural under the circumstances.

Maybe they could have their interview in a damp cellar. It would beat the heat. Why didn't he wear sensible loose clothes like everyone else? He would never get the hang of these foreigners.

"Was your master married? Any family? Next of kin?"

Otto shook his head, sipping the cool drink gratefully.

"No, just him."

Just once, Michael wondered, couldn't he have an easy case where the deceased carried his will around with him, complete with postal addresses and an inventory of his property?

"Anyone he was particularly close to, at all?"

Otto's head continued its gentle oscillation.

"No, just me."

He blushed at Michael's appraising gaze.

"No, no, not like that. He didn't really have anyone, any interests apart from his work, his business, he always had so much to do."

Another complication.

"Business?"

"He ran a successful printing and publishing house, set it up himself from scratch. It's going really well – not many printers will handle technical mathematical work with diagrams and everything. We're going to... we *were* going to..."

Otto broke off.

It hadn't really sunk in yet. Poor young guy, maybe tying yourself to a patron wasn't always for the best. Now, worth going for the million-to-one chance...

"I'm sorry, sir. I know it's difficult, but I have to ask these questions. I don't suppose that he carried a Last Will and Testament? Many travellers do, you know, what with one thing and another."

"No, not that I'm aware of. I suppose I'll have to go through his bags... I..."

"Don't worry, sir, I'll take his bags with me. You've been very helpful. I suggest you go and get some rest, it looks like you haven't slept for a while. Thanks again for your help."

Michael drained his cup and stood, adjusting his clothing and picking up the parchment. Next stop, the ambassador's office, see whom he should contact next at home. He would ask the steward to send the bags round to the office, there was no point in taking them now.

"Signor Benevento?"

Michael was already on the way to the door. He stopped, turned, and quirked an eyebrow.

"Yes?"

"You will get him, won't you? The man who murdered my master?"

July 10th

Signor Federico d'Urbano,
Chief Clerk to the Curia,
The Papal Chamberlain's Offices,
The Vatican

Federico,
I hope that you can help me with a tricky situation. I've got a dead astronomer on my hands, name (amongst others) of Johannes Regiomontanus. His assistant thinks he's been murdered, to do with a matter of academic and Church politics. The lads over here aren't as well up on these things as your people.

The whole affair is driving me crazy. Nobody seems to be able to give me a straight answer to anything. I've been speaking to this assistant (Otto, from Nuremburg) and a few university contacts of mine. A couple of the names that keep on coming up are some of the Greeks who came over when it all went wrong in the East, bringing all their books with them, so I thought of you – I'm sure you know people who are far better placed to look into some of these things, and I can trust you to tell me whatever you find.

JR was a protégé of old Bessarion, who died a few years back. It seems that B taught JR to read Greek and gave him a few astronomy books to crack on with translating. (There was another guy involved as well, but I'm trying to keep this short. I've learned far more about astronomy and astronomers these last few days than I ever wanted to.) Most importantly, he – JR – was going to prepare a new Latin edition and commentary of a classic book by an Ancient Greek, or Ancient Egyptian, called Ptolemy. It's called the *Syntaxis* or the *Almagest* depending on who you speak to. (Told you nothing here was straightforward.) Yes, these details are important, bear with me.

Now we introduce our Prime Suspect, another one of the Greek exiles, one George of Trebizond – the same man who wrote that *Intro to Latin Grammar* that's on all the young priests' desks these days. He was secretary to His late Holiness Nicholas V. George was, once upon a time, another one of Bessarion's circle. Now, G had written his own translation of this *Almagest* into Latin, with his own commentary. In it, he completely trashes the work of another old-time Greek/Egyptian called Theon. Unfortunately, Theon is one of B's favourite authors, so he waves goodbye to G. It doesn't stop there. G and B argue in print about philosophy (one of them thinks we should all learn Plato, the other thinks no, Aristotle is the bee's knees, and I don't particularly care which was which). Then B gets nasty (you know what he was like) and says, "Not only have you got the wrong ideas, Georgie my lad, but your translations are full of mistakes – look, here's a little list of a couple of hundred I spotted in just one of your books."

So, long story short: George of Trebizond ends up hating Bessarion and all his works, and all his hangers-on (including the new blue-eyed boy, young JR). I hope you're following all these names (it took me a few goes), because this is where it starts to get interesting.

Let me bring things up to date. JR beavers away, preparing his new version of this old book. He's staying in Bessarion's palazzo here in Rome, and finishes the thing in '62, but doesn't try very hard to get it published. There's no particular rush and he'd like to compare it to some other manuscripts in B's collection. B goes off to be the Pope's ambassador to Venice, and JR goes back over the Alps; he picks up various jobs, writes some maths books (about triangles! Did you know you can buy whole books about triangles?) and eventually ends up back home in Franconia, wherever that is.

That's where our friend Otto met him. They set up an observatory, a printing press, start... well, observing and printing. JR's still brushing up his *Epitome of the Almagest* while he rattles off a few other books, and he prints a list of all the stuff he's going to publish over the next few years, to show off at the Book Fairs and send to universities and suchlike. Quite a keen businessman, by all accounts. I think Otto's sincere, if a bit naïve, by the way, I don't suspect any foul play there. In this list, JR highlights his new book and says that (I'm quoting from the copy Otto gave me) "he will clearly show George of Trebizond's commentary to be utterly worthless and his translation full of errors." But what does he care? He's got no contact with George, and anyway, he's hundreds of miles away.

Then His current Holiness gets all in a bother over reforming the calendar. My astronomer friends got very excited about this, but I'll freely admit I didn't understand a word of it. Heavens, Federico, the conversations I've had this past week, my head hasn't stopped spinning! His Holiness summons all the best astronomers from all over Christendom, and of course that includes JR, who duly turns up, starts attending meetings, talking to other folk, making his presence known. Everyone's asking him when his book is coming out, the universities and booksellers are buzzing... and three weeks later, Johannes Regiomontanus complains of stomach pains and keels over, stone dead.

Now, it could be the plague that's been going around since the floods. It could be just one of those things. But... it does seem an odd coincidence, wouldn't you say? Something isn't right. I'd be very, very grateful if you could turn up anything on this George, see what he's been up to, what his reputation is, you know the drill. Sorry to have bored you with all the background, but something might be important here.

Let me know what you find. I might even stand you a drink sometime. Hope all's well with you.

Best wishes, Michael B.

July 20th

Papal Chamberlain's Offices

Michael took another paper from his desk. The English ambassador couldn't help locate any relatives for the merchant found washed up in the harbour. Normally this wasn't any of Michael's concern, but this one had been commissioned to take some private – and secret – letters to London. Another day, another mess.

"Signor Benevento? There's a man to see you, sir, a Signor Nurnbergensis."

What could it be now? He turned in his chair, looking past the junior clerk who doubled as receptionist to see… could that be Otto? Yes, although he made a very different impression to the fish-out-of water he had met two weeks ago. He was wearing a fashionable long tunic now, and he'd even had a haircut. He looked quite the young man about town. At least, insofar as an academic could carry off that look. Michael stood, smoothing out his own expensive clothes.

"Otto! I was going to get in contact with you, but there are still a few loose ends to tie up. I never get the time to deal with anything thoroughly, we're pretty short-staffed here. Sorry, sorry. Please, have a seat – just move my hat off the chair and put it... somewhere, will you? I've got some news, my Franconian friend."

At least he had learned something useful from this case – Franconia was just beyond the Danube, and its chief city was Nuremburg.

Otto picked up the gloriously new sun-hat, the very latest fashion, and looked around for a free surface. The office was tiny, and almost completely filled by the desk, piled high with papers, and the chairs. Sitting, he balanced the hat on his knee. He looked much more relaxed than during the interviews they'd had, but that was only natural.

"You're looking well. I hope the Papal staff have been treating you well?"

"Oh, better than well, sir. You might not have heard – I've been appointed to the council on calendar reform in my late master's place, given that I know pretty much everything that he was working on. Hence the new threads."

He gestured downwards vaguely.

"And Regiomontanus' belongings reached you safely?"

They had been signed over to the Nuremburger for safe keeping and transport back north of the Alps until a will was found.

"Yes, thank you, quite safely. There are some fascinating works in progress and letters. He was revising the *Epitome* right up until his last breath – he's found a glaring error in Ptolemy, about the distance between the moon and the Earth, and it's really going to set the cat amongst the pigeons! We're going to need a whole new system of astronomy..."

Michael held up his hand. He knew that as soon as someone mentioned Ptolemy, it was time to change the subject.

"I'm sure it will. Can we talk about George of Trebizond instead?"

"Yes, yes, of course. That's why I'm here. It's just all a bit overwhelming."

"Hang on a moment while I find the right piece of paper. Here we are. So, we've ascertained that George is currently in Rome. He was kicked out some time ago, but it was only a temporary expulsion and it lapsed two years back. He's living in a cheap apartment, months behind on his rent. He spends his time dodging his creditors and the various churchmen who bear him grudges for all sorts of reasons, long story. However, he's in his eighties now, and not really up to running around poisoning his rivals."

Otto nodded.

"I never thought that the old man would have done it himself."

"Wait, you're getting ahead of me! George has got two sons, David and Christopher, both of them in their forties. They live near him, scraping a living doing translation, interpreting, that sort of thing. Showing tourists and pilgrims around, finding them places to stay, robbing them blind. Unpleasant characters, both of them, but not actual criminals as far as we know. David's brother-in-law is an apothecary, so he'd have access to all sorts of nasty stuff."

"Well, there you are! It must have been him!"

Otto shifted in his chair, and the brand new hat, the hat that would be the talk of the Chamberlain's Office, the hat that had cost a good week's wages, slipped onto the dusty, horrible, floor. Michael winced, but his visitor was oblivious.

"Can't you get hold of him, make him talk?"

"Make him talk? I don't know what people are like in your part of the world, Otto, but this is *Rome*. We do things by the book – in fact, we write the book."

He was quite pleased with this, repeating it to himself under his breath and making a note to drop it into conversation when he met Federico for that drink.

"Anyway, he had the means to commit a murder, and he had the opportunity, but what's the motive? Why would he kill one of his father's rivals to get back at a cardinal who died four years ago? Or about a book that hasn't even been published?"

Otto spoke quietly but firmly. He did appear to have grown up a lot, very quickly, but extraordinary circumstances have extraordinary effects.

"I don't think it was anything to do with Bessarion, I think that's a dead end. George is poor, very poor, you say. The man who was once a Papal secretary! He's reinvented himself, given up the philosophy and the astronomy and come up with a best seller: his grammar textbook. He must finally be earning some money to pay his debts. What's the worst possible thing that could happen? He loses his reputation *again*, my master's book comes out and takes him apart line by line – it's really very funny to see his destruction in the manuscript. People laugh at old George again, his book falls out of vogue, he's back into the gutter. Wouldn't you slip poison into someone's food to prevent that?"

Michael shook his head.

"I don't know. It doesn't add up."

And he only had Otto's word for it that stomach pains were even involved.

"But I know that His Holiness himself is interested in sorting things out, so I'll speak to some of my colleagues in the Holy Office and see whether we can find anything incriminating about David, have him followed for a while, retrace his movements recently. You'll be at the guest quarters?"

Otto nodded, standing. As he pushed the chair back, it caught against the brim of the hat.

"Please let me know the moment you hear anything, sir. I'm anxious to get everything settled here. There will be need for new management back at the printing works as soon as I can get away, and I wouldn't want to miss bringing this criminal to justice."

He smiled, and bowed a little courtier's bow. Michael wondered who had been coaching him.

The overworked investigator nodded vaguely, half-rising from his seat, trying not to think about his lovely new, crushed, sun-hat. As the curtain that served as a door swished shut, he shook his head to clear his thoughts.

He dipped his quill into the ink-pot, and began another letter to Federico.

Notes on Chapter Ten

Regionmontanus

Fictional characters:

Michael Benevento, Otto (again!), Federico d'Urbano. I have given names to the two sons of George of Trebizond

All of the characters in this chapter, apart from Regiomontanus himself and the family of George of Trebizond, are my inventions. We know that Regiomontanus died young while in Rome on Papal business, and foul play was alleged almost instantly. It is much more likely that the plague in Rome at the time was responsible, but how can an author pass up the opportunity for a murder mystery? In reality, Regiomontanus' printing press was taken over by his business partner.

The convoluted background to his story is, however, entirely true. Regiomontanus was a child prodigy. His talent was spotted by the greatest astronomer of the time, Peurbach, and the two worked together producing textbooks and commentaries as well as original theories. He soon saw the advantage of the new printing technology in overcoming the recurring problems of copying errors. He also saw that dedicated scientific presses were necessary, not least because general printers were unwilling to buy expensive type to set up pages of numerals.

Regiomontanus was famous enough during his own lifetime for rumours to spread of his incredible abilities (he was meant to be able to construct flying automata). There is a convincing argument that Dürer's enigmatic engraving *Melencolia §1* is a tribute to him. He was also the first person to have noticed Ptolemy's cover-up of the lunar distances – he wrote to a friend that this was a "jagged tooth that tore a hole in astronomy," but died before he could attempt a solution.

11

Ambition

AD 1492

In which we learn that when, in fourteen-hundred and ninety-two, Columbus sailed the ocean blue, he did so despite his almost complete ignorance of all the relevant astronomical and geographical facts, and that his Voyage of Discovery owed more to the whim of a King than to any sound planning.

A small inn, in southern Spain

Rodrigo sat forward on the edge of the uncomfortable bed, which half-filled the tiny bedroom. Hardly an ideal room for their practice interview, but you got what you paid for, and this had been free. His client had been given a Letter of Retainer by their Majesties Ferdinand and Isabella, which entitled him to a small salary and to free board and lodging at any commercial premises in the kingdom. It had taken some time to persuade the stubborn sailor that a good use of some of that salary would be to engage Rodrigo de Escobedo, up-and-coming lawyer, but eventually he had seen sense. After all, he'd been approaching princes and monarchs by himself with his plans for the last four years to no effect, and professional advice was obviously required.

Sometimes Rodrigo wondered whether their Majesties had granted the Retainer simply to buy some peace. Señor Cristóbal Cólon could be a little overwhelming once he got going.

"Right, Chris, whenever you're ready."

Chris was on his feet, adjusting his tunic, looking down at the floor. He launched into his speech.

"By your Majesties' Leave, I wish to present the developments in my plan to open the trade of the East with all of its luxuries and wealth to your glorious Majesties' merchants by a direct sea-route sailing westwards from the Canary Islands as I have already explained to you in the past, but now I have discovered some more facts which change the situation considerably and enable me to..."

"Stop, stop! You're gabbling, man. Short, punchy sentences. Remember? You won't have much time to make your pitch, and you want every word to count. You're a salesman, you've been in business for years. You're selling an idea now. And look up, for the Lord's sake! Just because they're royal doesn't mean you have to look at the floor."

The middle-aged, heavyset, man sighed, running a hand through his thinning hair. He looked up and pulled a face.

"You're right, Rod. It's just that I know this is my last chance, I've had two audiences already and they've told me each time that I need more evidence."

I'll give him this, thought Rodrigo, *he tries his best and he takes advice well. He would learn. The only question was, would he remember his lessons tomorrow? He'd been drilling him for weeks and look where they'd got. At least that Italian accent was fading. He supposed that was what living with a good Spanish girl did for you. Better not try to play the 'wife and kid to support' card in the interview, though. She was not actually his wife, the kid was by another woman anyway, and they hardly wanted their oh-so-devout Majesties to find that out.*

"Cheer up! They gave you the Retainer, so you know that they're interested. You just need to say the right word at the right time. This is most definitely the right time. We'll get you that funding, on your terms, or I'm not your lawyer."

Let him interpret that how he may. It was his last chance, too, with this client. If Chris got it, Rod would push hard – he would have to give him a permanent job, maybe legal advisor to the fleet? There would be such a lot of paperwork to get through, not to mention a cut of the potential profits. They just needed to buy him time, keep them listening long enough to get them interested in the possible returns on a minimal investment.

"Right Chris, from the top. Remember: punchy. Start by flattering them, as we agreed, tie it in to their great victories. Just like we discussed over dinner last night."

Cristóbal nodded curtly.

"Yes, I remember. Here we go."

He cleared his throat loudly.

"Your Majesties. I have come with more details of my plan to make your kingdom wealthy beyond compare. You have reconquered all of Spain's ancient lands, and now you may reconquer her ancient wealth."

Rodrigo smiled. He thought that was a good line, even if Spain had never been particularly wealthy, and was pleased that his client had remembered it correctly. He nodded his approval vigorously.

Why else would they establish their Court in the Moorish stronghold itself, if they didn't want to make the point about their military victories? Even if it is inconvenient for everyone to travel all the way down here to Córdoba, where the decent inns are all full of courtiers. I suppose that being royal means never worrying about the inconvenience you cause to others.

What now? He's stalled already. This is never going to work. If only he'd let me go in and speak on his behalf, it's what I'm trained to do! But he's right, I couldn't answer the technical questions as convincingly as he could, he's certainly read up on his stuff.

"Sorry, Rod. Should I mention the new facts now? Or recap the old ones?"

"I thought we'd gone over this. Straight for the new stuff, it plays to their religious sensibilities. Then be ready to defend yourself against hostile questions from their advisors, I doubt their Majesties will actually ask you anything directly. Don't bring up any of the things that you've mentioned before, until you're asked."

"Right, yes. So, flattery first, then the new material, then be prepared for questions. I'm sorry to be so slow, it's just so different from my natural environment, and my Spanish still isn't quite as fluent as I'd like."

"Nonsense, man, it's fine. You're a traveller, you'll adapt quickly enough. Think of it as being like an unfamiliar marketplace, you'll soon learn your way around."

"From the beginning, then?"

"No, start with the new material. Then I'll cross-examine you in the guise of one of the so-called scholars. Deep breath, slow but clear. You need to speak slower than you think, nerves will make you speed up, even here and doubly so at Court."

A pause. This time, he didn't look down at the floor. Genuine progress!

"I have consulted many texts, both ancient and modern. I have found sure authority for my project which cannot be rejected. The Bible itself confirms my geographical suggestions. In the second book of Esdras, in the sixth chapter, we read that 'the waters are gathered in the seventh part of the earth: six parts hast thou dried up.' So we have Divine authority that the world is six-sevenths land and one-seventh water, and so the Ocean Sea must cover at the most one-seventh of the globe."

The lawyer applauded.

"Well done, Chris, that was superb! Short, pithy, to the point and obviously relevant to your cause. I don't see how they can argue with that. You'll probably get applause like this in Court, you know how their Majesties adore scriptural passages. Now we'll try the questioning. Remember, you'll only be in there for ten minutes or so at most, so if you can't give a good solid answer, or if things are going wrong, return to that Biblical passage."

"I think I'm ready."

Rodrigo sat up, and assumed a different voice, high-pitched and querulous, for his role as an elderly and sceptical astronomer or mathematician.

"So, Señor Cólon, your submission to this Court shows that the circumference of the Earth is twenty thousand miles, therefore one-seventh would be roughly three thousand miles. A long voyage, but hardly impossible. On what do you base your figure for the circumference?"

"Sir, it is from the late Cardinal d'Ailly's incomparable book *Imago Mundi*, used as an authority by many geographers and theologians."

"How will you accurately establish your position when you are so far away from land?"

"Sir, I have the very newest instruments for the recently-developed art of Celestial Navigation, purchased at my own expense in Portugal, and I have the latest and most accurate tables of planetary positions, prepared by the renowned astronomer Regiomontanus, sometime consultant on calendar reform to the Pope."

Rod nodded. Stressing Papal connections was another good fallback that they had agreed upon.

"Why should we finance this expedition?"

"Sir, the sea route to the East will be faster and cheaper than the Silk Road. Sailing Westwards, we will reach the archipelago of Cipangu, the chain of islands to the east of China, directly and avoid the long and treacherous coastal route being pioneered by the Portuguese explorer Dias."

"And what do you require?"

"An exploratory fleet of three ships, fully crewed. Your Majesties would only have to provide half of the money as I have found private backers already."

"That seems very encouraging. Your reward?"

"The rank of Admiral, and a share of all wealth generated from the trading route, the details to be decided at your Majesties' pleasure."

Rod smiled, and leaned back, resuming his natural voice.

"Excellent, Chris, short and to the point! Not like those rambling monologues of a few weeks ago, eh?"

His client nodded happily.

"You're worth every *maravedi*, Rod. I couldn't do this without you. I think I'm as ready as I'll ever be for tomorrow. Probably time to turn in for the night."

"Sure. I'll see you in the morning. What could possibly go wrong? Good night, Chris, and dream of titles and treasure."

Rodrigo closed his eyes and slumped back against the bench, wishing that he could become invisible. The throne room of the Alcázar castle was lavishly decorated with tapestry hangings. Couldn't he simply blend in to one of the hunting scenes somehow? This was a nightmare, and the calamitous end to what had been a profitable and enjoyable job. Sitting along one wall with the other petitioners awaiting their audience, he watched poor Chris stumble from disaster to disaster.

It wasn't his fault, the lawyer told himself. He had no idea that the questioning would be so very hostile and personal. How could he avoid his name being linked to this fiasco?

It had started relatively well, although it was annoying that King Ferdinand was dealing with other matters. He was by far the more business-minded of their Majesties. Cólon had launched into a slightly expanded version of his introductory praises and delivered his Biblical passage very clearly. Then Isabella's chaplain had started the questioning, backed up by a spectacularly wizened academic of some kind.

"How had this common sailor the nerve to quote chapter and verse when he was living openly in sin with a girl twenty years his junior?"

Ouch. That was obviously intended to turn Isabella against Chris. Her Majesty had rejected plenty of potential suitors in her youth, purely on the somewhat odd grounds that a husband and his wife should be of similar ages. Chris's indignant answer – "Only fifteen years, sir," – hadn't helped, and it had gone downhill from there with continued questioning about his personal morality. At least now the chaplain had resumed his seat, and the scholar had taken over. This was the time for recovery, but Chris was looking so riled and angry that the prospects weren't good.

"Señor Cólon, in the documents you presented with your petition you make use of an estimate global circumference of some twenty thousand miles. This is taken from..."

"Cardinal d'Ailly's..."

"*Imago Mundi*. I know. It's an interesting little book, if somewhat outdated. Are you aware that the section on the size of the Earth is taken directly from Alfraganus, a Muslim?"

Isabella's eyes narrowed and she clutched the arms of her throne.

"Yes, sir, but he was an excellent mathematician and..."

"And he calculated in Arabic miles, not in Spanish or Italian miles. You will of course know that these are half as long again as our measurements, and so the true circumference is closer to thirty thousand of our miles? A ship couldn't carry that much food and water."

"I... that is..."

"Come on, man, who taught you astronomy and geography anyway?"

"I... taught myself, sir. From the books."

There was a politely restrained sniggering from the assembled Court. Cólon was indignant.

"But I can assure you..."

"You'd simply sail westwards, and starve after three thousand miles. That's in *our* miles, you understand. Now, you claim that Cipangu is an archipelago which stretches further eastwards than we currently know, making your proposed landfall easier. On whose authority?"

"I... there was a learned Italian, sir, one Toscanelli, who..."

"Ah! A university astronomer, or an explorer, no doubt?"

"No, sir, a medical doctor, but..."

More laughter from the Court; some of the courtiers pointed at a smug looking fellow, obviously the Royal Physician, who rolled his eyes ostentatiously.

"Turning to practical details. These new astronomical instruments of yours, you are experienced in using them?"

"I was fully trained in their use, sir."

"How many times have you used to them to navigate on the open ocean, out of sight of land?"

"Well, I haven't... there hasn't been..."

"I see. You have found private finance for half of the costs your expedition, so why cannot you recruit more backers for your damn-fool project and save their Majesties expense? Our finances are depleted after the war."

"It's... I thought..."

The scholar turned to the Queen.

"This... gentleman... has failed to secure funding for his project from Portugal, your Majesty, from his native Genoa, from Venice and possibly from other Italian states although our intelligence of such matters is not certain. I recommend dismissal of his petition."

The priest alongside him was nodding vigorously, muttering something about the need to set a high moral tone for the newly reunited kingdom, not being seen to patronise mercenary adventurers and fornicators.

Isabella considered for a few seconds that seemed to Rodrigo like hours.

"Petition dismissed. The petitioner may approach Us again next year if he addresses the issues raised. Next!"

Then suddenly he was being ushered out of the Court, along corridors, through the guardhouse, and out on to the street, next to his ashen-faced and silent client.

They drank wine in the rowdy lunchtime tavern in silence, each alone with his thoughts and unwilling to shout above the noise. Eventually, Chris could stand it no more. He left his goblet on the table, and clapped his arm around the young lawyer's shoulders, gesturing towards the door with his head. Then they had reclaimed his baggage mule from the inn's stable and checked that all his goods were still there.

As they approached the city wall, the two men walking alongside the mule, Rodrigo took a deep breath.

"Where next, sir?"

"Let's just get out of town, find a country inn somewhere, sit down and rethink our strategy."

He suddenly smiled.

"Did I tell you that Bartholomew has made contacts with some English courtiers, who think that their King Henry might be interested? Not such a bad people, the English. I spent some time in Ireland once, which is pretty much the same thing, I understand."

He was irrepressible. Rodrigo felt that he should have known that; Chris had been on this mission, his personal crusade, for something like five or six years. He would probably keep going until he was dead. He must have a thick skin, or he would have given up ages ago. What use was there for a Spanish lawyer in London, though? Maybe he would need someone to act as his representative in Spain to look after his interests – he did still have that Retainer from the Crown, and permission to re-petition was better than a once-and-for-all refusal. Rodrigo smiled, realising that his client's relentless optimism was starting to colour his own views.

"But there's one thing I would know, sir. The details about the geography that the Queen's astronomer quoted, the different length of mile – is there anything in that?"

Cólon stopped. He turned to face his advisor, one of the few men he could trust.

"There... might be. It's all a matter of translation. But he, and all his kind, just cite their damned authorities, they don't know these things, they simply copy what their professors said. I taught myself everything, and not just from books, you understand, but from the world. They've never even been on a boat, let alone on the Ocean Sea. The things I've seen! Trust me, Rod, I feel it in my bones. Sailing westwards simply *has* to be possible."

His eyes glistened, and his voice was careful, subdued, holding back some strong emotion. Anger? Sadness? Conviction? Yes, conviction. Rodrigo was vaguely reminded of an incident in his youth, when he had forgotten part of the catechism; his old parish priest had tried to reason with him, to get him to think things through for himself, before giving up and simply beating him.

"I wish that chaplain had kept Beatriz out of things, though. You're my lawyer, isn't there such a thing as slander, or libel, or whatever you call it?"

"I suppose I could look into that."

No chance. They weren't married, and there was no point in

bringing such an action, but now wasn't the time to explain that.

"She'll be so disappointed. I was so confident that this time they would finally see sense. It's not as if I'm asking for very much upfront cash. Do I seem unreasonable to you, Rod?"

Rod shrugged, contorting his face into a gesture which could have meant almost anything. He'd a lot of practice in that.

"Señor Cólon! Señor Cristóbal Cólon!"

A shout – a liveried guard on horseback, riding them down, shouting and waving, scattering people out of the way before him. Oh Lord, what now? They were about to be arrested for fraud, or wasting royal time, or fornication, or preaching Arabic doctrines? Rodrigo's hand fell naturally to the dagger in his belt, before he realised how stupid resisting arrest would be. This wasn't a tussle with some wayside brigands. He opened his hand, and turned his gesture into a low bow as the Palace Guardsman jumped from his horse. Chris simply stared, his closed lips working a little. Good idea; maybe prayer was the best strategy now.

Rod held his position, gesturing with his hand for his companion also to bow. It always paid to be polite to the authorities. The sailor caught the hint, and adopted the same stance as his lawyer.

"I'm Columbo the Italian, called Cólon in this land."

The guardsman returned the bow, with a flourish of his cape. At least he hadn't drawn his sword. Obviously he had realised that they would come quietly. Very good! Their diplomacy was working at last.

"Sir, you and your... advisor are to return to the Palace immediately. His Majesty King Ferdinand, having heard the details of your proposal at second-hand, wishes to speak to you forthwith and to discuss the detailed financial terms and rewards relating to your proposed expeditions. I am at your disposal – Admiral Cólon."

As the guard saluted, and two more soldiers caught up with him, leading horses, Rodrigo blinked a few times, slowly unbending himself from the bow. Was this even vaguely plausible? It was the sort of improbable event which would be laughed out court if presented in evidence; but, yes, it was really happening, and the guards were smiling. He recovered his composure enough to exchange a delighted smile with his client. Legal Secretary to an Admiral... and possibly, soon, to a colonial governor and a millionaire. He had known it would all come right in the end. Not a bad day's work after all. Time to earn his keep in the negotiations...

Notes on Chapter Eleven

Columbus

Fictional characters:

None

Columbus went under many names to make himself more pronounceable as he travelled the Mediterranean in search of funding. He did engage a lawyer called Rodrigo de Escobedo to handle all the paperwork relating to the expedition and to his claims, but it is entirely possible that this only happened after he was (finally) successful in attempting funding.

The actual proceedings of the 1492 interview with Isabella are lost to us, but the chapter's reconstruction fits with all the known facts and the errors that he had made in his calculations and preparations. It is a source of constant irritation to historians of science that the general view of Columbus is that he was rejected as a fool for thinking that the world was round. As the reader will be well aware by now, the world had been known to be round since the time of the Ancient Greeks – people laughed at Columbus because his theories vastly underestimated its size.

The too-good-to-be-true sudden message from Ferdinand as Columbus was leaving town is, in fact, true. Ferdinand was tempted by the potentially huge return for a very small initial outlay (outfitting a "fleet" of three small ships was neither here nor there compared to other budgetary concerns at the time).

It is well known that Columbus died thinking that he had sailed nearly all the way around the world (hence "the West Indies"). His navigation throughout the voyage was appalling. If America had not been there, he would simply have run out of food and drinking water after roughly 3,000 miles, and never been heard of again.

12

Doubt

AD 1521

In which Nicholas Copernicus appears, the man who overthrew Ptolemy and put the Sun in the centre of the Universe, and we find that, rather than the revolutionary he's sometimes painted to be, he was a cautious and careful man feeling his way through a complex maze of conflicting ideas.

Frombork Cathedral

To His Grace the Bishop of Warmia,

Next time I am seconded to help the Government with economic administration, please advise me whether I am likely to end up in the middle of a castle under siege. The situation was so bad that I was pressed into service as a temporary General Officer of the Warmian Army, defending our small country against the Teutonic Knights – imagine that! I suspect that I looked the part far more than I felt it, and if I managed to inspire any of the troops then my job was done sufficiently well. Perhaps my look of bewilderment was taken for one of stern reproof – the soldiers behaved magnificently under trying conditions. Reports of my heroism and strategic brilliance are certainly exaggerated.

It's good to be home, although a home that I scarcely recognise. I hope that I am correct in assuming that the Diocese will help to pay for the repair and refurbishment of my study/office in the

Northwest Tower. Unless things have changed vastly since my last term of office as Cathedral Treasurer, I imagine that the Chapter's 'extraordinary events' fund would run to about a third of the finances required, and I shall pay for the rest from my stipend. Luckily my house itself was untouched by Albrecht's infamous raid, which by all accounts was launched more out of pique and the chance of easy pickings than for any more serious strategic point. I sincerely pray that the recent peace treaty with the Knights (how such self-serving warmongers can call themselves an order of chivalry, let alone run their own small state, is beyond me) will put an end to this petty squabbling and firmly establish Warmia as a staunch ally of our powerful neighbour Poland. Our continued prosperity, in Frombork and beyond, depends upon it, and upon attracting more farmers to work the land abandoned in the conflict.

Your Grace will excuse me – I am still thinking like the economist I have been for the past couple of years. Rest assured that I have no intention of reforming the coinage of the city, or indeed of pressing for high rank in the militia. I hope to sit peacefully in my study (at least, once the wall has been sufficiently repaired) and devote myself once again to medical and legal problems on your behalf. Has anything of note happened in my absence? I can't wait until the next full meeting of the Cathedral Chapter to find out, and everyone is so busy rebuilding and rearming that I hesitate to disturb them. I should be very grateful of a full report so that I make no errors in the conduct of our pending ecclesiastical cases (the file is sitting on my desk as I write this).

I can also reassure you that I fully consider Frombork, and my canonical duties here, to be where I belong. In your letter, you asked whether my time at Olsztyn has given me a taste for high administration and the city life, whether I would be content with our sleepy little town (well, sleepy until roused by marauding

Teutons). The answer is that I am a thousand times content! Your Grace will recall that I started my university career in Krakow, and pursued my medical and legal studies in the venerable institutions of Bologna and Padua, that I have seen the glorious city of Rome itself and learned its language. I have travelled more than enough for any man, and am looking forward to spending the rest of my days in a comfortable house in a small town, doing work which is pleasing to man and to God. Warmia is a wonderful place to live and to work – now that the young upstart Luther seems determined to set the world on fire, and has given every rascal in Christendom an excuse to start fighting, nowhere seems particularly safe. I had as well make my stand here as anywhere else. Have no worries on that score.

Unfortunately, the raid also damaged my astronomical instruments. These will prove more difficult to replace. I know that I have just returned, but I beg your Grace's permission to leave again for Thorn, or possibly Krakow, to commission replacements from the artificers there. The trip should take no longer than two or three weeks. Much though I love Frombork, explaining the precise techniques required to produce an armillary sphere to the local blacksmith would be an exercise in frustration. I speak Polish, Latin, Greek, Italian and German, and still cannot communicate with the local tradesmen... I must tell you that although my astronomical studies are strictly a hobby, your kind toleration of the time that they require has been a constant source of pleasure and gratitude to me.

While I am on the subject, thank you for your kind remarks on my little notebook of thoughts concerning the various hypotheses about the universe. One day I might, indeed, seek to publish them as you urge! As it is, I have been satisfied for the past few years with circulating the manuscript amongst my closest friends and colleagues, as each fresh pair of eyes brings something new for me to consider. I found your comments on the nature of the analogy between God and the Sun most interesting, and have been thinking about them constantly since finding your letter upon my return. (Please hold on to the booklet itself until I can collect it from you personally, as I only have a couple of copies.)

Firstly, you are right that my new understanding of the universe would require major changes to the noble and useful art of astrological prediction. Although this has not reached the technical perfection of our descriptions of the nightly planetary motions, due to the number of other influences acting upon us, it would be a foolish man who discarded such an important tool. I must disappoint you, though. I am not the man to carry out such a reform, since the influence of the planets and stars upon us humans is something I have never studied in any depth. I cannot even begin to say how the new position of the Sun would affect the subtle threads binding the disparate parts of the universe together into one continuous and interconnected Creation.

You also touch on a subject which I have considered deeply. The nature and definition of "perfection" in the physical world is almost as dear to me as the perfection of the soul itself, one of our great endeavours. God has created His system carefully, and for our moral edification. Circles are a sign of endless perfection, as the philosopher Aristotle, his commentator Averroës, and his more modern disciple St Thomas Aquinas agree. (I heard recently that certain Dominicans are pressing for St Thomas' formal recognition as a Doctor of the Church. I wish them luck in persuading their Order to back this noble cause!) It is obvious to me, as well as to you, that the universe must be formed from a combination of circles and spheres, for the good theological reasons which you mention in your letter. Various astronomers have attempted to construct such a combination, to trace the Lord's handiwork, over the years – the most successful attempt is undoubtedly that of

Purbachius and his student Regiomontanus, whom I admire without reservation. Their technical mathematics is simply breathtaking: I have dabbled enough with the basics of trigonometry to appreciate masterly work when I see it, as only a man who has struggled with a paintbrush or an etching knife but has little talent can truly appreciate the work of a Dürer (who has himself produced a masterly print celebrating the memory of Regiomontanus!)

I have no quarrels with this intricate and beautiful framework of whirling circles, wheels attached to epicycles, all circling – all circling what? Well, since Ptolemy we've known that they don't all circle the same point, as their eccentres are displaced from the centre of the universe in different directions. I stressed in my notebook the good physical reason for thinking that the Sun is the body closest to the universe's centre – Mercury and Venus never wander far from that luminary, and the most obvious explanation is that they are circling around it at no great distance, and within the path traced by the Earth's own orbit. But is that sufficient to dispute the consensus of two thousand years? (Yes, others have thought that the Earth might rotate, or move somehow, but none have really argued their case.)

In this uncertain world, we can look to Divine principles to guide our understanding of matters which reach beyond our imaginations and knowledge. In the case of the whole system of the universe, our Earth is changing, mortal, fallen (although redeemed by Christ's mercy). The Sun is perfect, the source of all light and heat, the wellspring of life itself. Which is more fitting to be at the centre? This is why I was so encouraged to read your kind thoughts on the moral status of heliocentricity, since if we take the created world to be an analogy for spiritual truth, as was urged by Augustine and the other Fathers, then the Sun is the visible analogy of the Lord Himself, around whom all things revolve, and without whom there would be nothing but darkness and the void.

What is left of our Earth? Are we attached to an epicycle, or is our orbit perfectly circular as befits the most perfect of the planets, which was visited by God Himself and on which his chosen creation dwells? Now we reach a thorny problem indeed. I have been wrestling with this of late, and if I present the alternatives to you, I hope that you will be able to offer some theological insight when we next have the opportunity for a long conversation.

When I wrote the notes which you have read, I believed that the Sun, as the most perfect body in the universe, must have the most perfect and sustainable motion, *viz.* no motion at all. It, rather than the Earth, must be truly stationary. While the other planets, including our own home, do not have the centres of their orbits precisely at the centre of the Sun, we make take that point to be the centre of the universe. This has the virtue of elegance and simplicity, but at what price? The Earth must be simply the same as the other wandering bodies, with a similarly complicated compound movement.

On mature consideration, and after much prayer, I find this hard to accept. The Earth is manifestly different from the other heavenly bodies in so many ways. It is composed of four elements rather than simply one, and is the home to all of us, chosen by the Lord to have dominion over the whole of Creation (which we must take in a broader context than simply this Earth). Surely, this must be reflected in the physical foundations of its motions; surely, our Earth must be more perfect than the other moving bodies. Thus we can represent its motion by that of a simple, perfect, circle – which in turns demands that the very centre of the Earth's orbit must be the very centre of the whole universe. The mathematics is difficult, but not insuperable, and I am confident that I can make this change of perspective and still recover the observations which I have made all my adult life, and indeed those of the published tables of planetary positions calculated by other means.

However, there is a consequence. If the centre of the universe is the same as the centre of the Earth's simple circle, then the Sun cannot be stationary. Indeed, it must be circling on a small cycle of its own. Where does this leave the theological notion of perfection, which surely is applicable to a higher degree to the Sun than to the Earth? I am at a loss, as it is manifestly obvious to me that we cannot have a system in which the Sun is at rest while simultaneously the Earth moves in a simple, perfect, circle around it. How shall I proceed?

I pray nightly on this matter, and am looking for hints and clues throughout Creation. I am happiest searching the Heavens, whether it is with my eyes or with my mind, but I always hear a voice whispering at the back of my head that I have overlooked something obvious, that there is a simple solution. It is no consolation that if there is, the greatest philosophical minds have failed to find it. This is not pride, or at least I hope that it is not, simply a recognition that I am the right person, living in the right age, to do work that will glorify God, however quietly.

I find myself out of my theological and spiritual depth, and would be truly grateful to hear your thoughts on the matter. Physically, I cannot think of any good reasons to distinguish between the two possibilities. The question nags away at me, stopping me from concentrating on more readily accessible problems. Sun at rest or moving? Earth on an epicycle or not? Which is more perfect? How does the universe represent the Lord's intentions? When I was briefly a General, I noticed that in the secular world as in the religious one, the soldiers' chief concerns were: who is in charge? what are our duties? Which is to say, what is at the centre of our endeavour? what paths do we follow? The Lord has shaped our minds so that they echo deeper themes which He will reveal to us in many different ways.

Perhaps you will tell me, in the manner of some of the Greek philosophers, that it doesn't matter, that the systems are equally good if they produce equal results. Perhaps these wheels are human inventions, a drawing scribbled across the sky (as some say of the constellations). I simply cannot accept that. Although I have the greatest respect for their authority, they were pagans who did not know the Truth of Christ, and therefore did not appreciate that Truth is one thing and endures from generation to generation. It is abhorrent to think that two different explanations of something could share in equal truth, an error of logic as much as of fact. Of course, this is one of the many places that Luther falls short of the mark. He suggests that the Church is a purely human institution and that therefore it is open to arbitrary change. As the visible Body of Christ, we are following divinely ordained patterns, and I see the same things happening whenever I see the night sky. The Lord is never ambiguous, and if we cannot immediately discern His purpose in any element of Creation, the fault lies with us for not sufficiently using our talents.

Of the pagans, Aristotle saw glimpses of the Truth (a fact recognised and celebrated by the Thomists). The accretions and corrections of the centuries have dulled the purity and brightness of his vision, and if my booklet, my thoughts, my astronomical work as a whole have done *anything*, it has been to re-establish the central importance of perfection as the explanatory concept underlying the physical, observable world. So if I can wrestle successfully with that problem, my work will advance hugely, and maybe, just maybe, I will find myself in a position to publish. Do you think that others will be interested in my researches, theories, and speculation? When I have had opportunity to speak to genuine academic mathematicians and astronomers, they have tended to reject me out of hand – I am not bitter about this, as they have not been able to see the detailed notes which I have been working up on the subject and they are not always fully conversant with the underlying religious issues. If my conjectures have captured anything of the Truth, it must be the corporate and corporeal guardian and apostle of Truth – our beloved Church – that will be my supporter and my strength, rather than the professors in their solitary towers. You may rest assured that I will certainly not shout my views to the world until I have had their theological underpinnings thoroughly investigated by those who know more about these things than me.

I know that you will sympathise with my need to strike while the iron is hot and to put down my thoughts before they escape me. It is a wrench to turn away from my philosophical hobby and from my genuinely pleasing work to deal with the day-to-day matters of getting my roof and wall fixed and fulfilling my social duties. (Everyone seems to expect a visit from the reluctant Hero of Olsztyn, all at once, whereas all I want to do is sleep for a week or so!) What did St Thomas recommend to the weary soul? Sleep, a bath, and a glass of wine!

Before I can turn to those, I have a far more pressing question for you, descending from the Heavens to the Earth with quite a bump. In all the trials and tribulations of the past few months, I see from your secretary's notes in the file that several of the Cathedral's tenants have fallen well behind in their rent. Do you wish me to prosecute the cases? Similarly, some of the payments due to us for the use of Diocesan craftsmen for private work are overdue. I would urge clemency, given the extraordinary circumstances. People need time to restore their confidence in their security and to rebuild their financial situations as well as their farmhouses! I will, as always, be guided by you.

It's so good to be home.

With the very warmest of wishes,

Your humble servant, Nicolaus Copernik

Notes on Chapter Twelve
Copernicus

Fictional characters:
None

Copernicus has been much studied, and it is unsurprising that he comes across very differently in the accounts of different historians. The recent trend has been to emphasise that he didn't think of himself as revolutionary, but rather as restoring coherence and purity to the circular ideals of the Greeks. The chapter follows this reading, with the consequences for the motion or immobility of the sun that plagued Copernicus for his entire astronomical career (his published works are ambiguous on the matter).

The biographical details in the chapter are correct. It is easy to put great thinkers into a one mental pigeonhole, and military campaigns into another, but the two sometimes collided (as in the case of Copernicus). It is also easy to forget how unsettled Europe was throughout history, and to underestimate the very real possibility that war would break out between neighbouring states, with consequent devastation of the border regions.

Copernicus was a loyal churchman, holding a succession of posts in cathedrals, and the Church supported his work at the time. We shall see how and why attitudes changed in subsequent chapters.

13

Despair

AD 1544

In which Rheticus, the Protestant academic first responsible for introducing the Catholic Copernicus' ideas to the wider world, expresses his tortured thoughts and defends some of the things he has thought, said, and done to his mentor, Philip Melanchthon, Martin Luther's right-hand man.

My dear Philip,

How hard can it be to say goodbye to someone whom one respects – no, loves? I seem to have been doing nothing else for the last few weeks, months, years – perhaps for all of my life. For your information, it doesn't get any easier.

It was thoughtful of you to enquire after me, the other day; Johan told me that you had been looking for me on your latest visit, wondering if I was in good health. I don't know whether I am or not. Certainly, I'm physically sound; but I can't speak for my mental faculties, nor for my spiritual health. You're the expert in that department, and so perhaps you can judge me in my absence, and count this as my confession. No, my *apologia*; I know that language is important to me, to you, that it's important to get these things right.

So this is to say goodbye to you, and to the German-speaking world. I've been as happy here as I think I could be anywhere, but the circumstances are insufferable now. I trust you enough to know that that you'll burn this letter, and that somehow you'll find a way to console Johan. Whatever else I might be, I'm certainly a coward. It's my duty to write to you – you were my boss, my inspiration, at Wittenberg, and the authorities here at Leipzig will be looking to steal one of your Bright Young Things to replace me, what with reliable astronomy professors being so thin on the ground these days – but what do I owe to him, other than to acknowledge how much he's helped me these last days? Do I love him, or have we merely been using each other? There's no way to untie the various knots, so I'm cutting them and moving off (to conquer new worlds? Hardly!) Please don't follow me, or have me followed; I'll probably be changing my name, yet again, and I'll be doing the best to change my life in so many ways.

Names are so important to you, you with the Greek pseudonym and the international reputation. You're the clever one. It's so elegant, the way that Luther appeals to the man in the pew, and you to the man in the pulpit, such a neat little trick – what would you two have done without each other? Perhaps that's blinded you, and you can no longer see how insignificant names are to me. Labels are only useful if they pick out the truth about the world, and I think that there *is* a truth about the world. I used to think that you agreed with me, but reading some of your more recent pamphlets has left me in doubt about your opinions. How can you say that we'll teach one interpretation of the Bible, of astronomy, of philosophy to some people, and another to the initiates? We're finally getting somewhere in understanding the Creation, and you want to turn us into a Mystery Cult? I know that you'll say that I've misunderstood you, and that I need to consider your arguments more carefully, and believe me, I'll try that; you'll find no more dedicated a purchaser of your tracts in Christendom, or at least in the Reformed part thereof! But such a slippery doctrine doesn't exactly encourage me to maintain my connections with my old colleagues.

Of course, you'll know, or have guessed, that that's not why I'm leaving in such a hurry, and in such... moral turmoil. No, no, it's nothing to do with Johan either, sweet boy that he is; I can find nothing in my heart to repent of in that regard. I have nothing but respect (that word again, I can't get away from it) for the way you've helped me to establish my career, and I can't think of anyone that has been more sympathetic to astronomy, and to predictive astrology. In these troubled times, it's reassuring that at least some theologians will stand up to defend both of these cornerstones of rational thought.

Philip, I wish I could speak to you face-to-face about all this, but I've made up my mind to go, and you're so very persuasive, and I can't risk it. It all comes back to that one initial impossibility; saying goodbye to what one loves. Perhaps it's because I never learned to do so when I was young – you know all about my father, and the difficulties that I still have in accepting that he's gone. I'm sure that nobody, at least nobody I'd like to meet, could claim that the execution of one of his parents could possibly be justified, but – my father was not only executed for a crime he didn't commit (that's common enough these days, as God is my witness) but for one that he couldn't commit simply because no such crime exists. Sorcery! Sorcery! By all the heavenly powers, my dear, I hope that you and Luther between you can divert the stream of Truth through the Augean stables. (Or do you not believe in Truth anymore?)

That's the key. Is it easier to take your leave of a person, or of an idea? What is the proper realm of love? Now I'm treading on your professional toes as a theologian, no doubt; sometimes I wish I'd gone in for the Church after all, at least I'd be able to hold forth in public without being told off for having ideas above my station. (No, no, not by you, but Leipzig is more jealous of the traditional lines of demarcation than Wittenberg ever was.) Or perhaps I'm getting muddled, since both became one in my mind. Who needs all those old personifications when there's a real flesh-and-blood person who embodies everything that you believe in? Is it blasphemous to talk of the Master like that?

I know you didn't like it when I called him that in my little book, but that's who he is. Was. *My* Master: and I, his messenger, his student, his servant. His John the Baptist (go on, rend your clothes, that would be a suitably Biblical reaction for you, except for the fact that you're far too urbane and well-tailored). Let the world remember him as by that title, or simply as Nicholas Copernicus (or however else you choose to Latinise his barbarous name, Kopernigkgkg or whatever it may be), and let them forget me, since we have betrayed each other. It's not that I can simply leave Leipzig; I have to abandon astronomy, to expunge it from my mind, to forget all that I know of it though it destroy me to do so. It's too painful. Astronomy will henceforth be a synonym for Treason in my mind. And it's all because of you, Philip, or at any rate because of what you've made of me. *Odi et amo – excrucior.*

When I left you on my travels, left the seeming certainties of the world I knew, I hardly suspected that there could be any such person as my Master. Astronomy everywhere was a mess, and those who clamoured most loudly for reform were the people who wouldn't recognise it if they saw it. The early days of my sabbatical simply confirmed that one place, one method, was as bad as another. We were all scrabbling about trying to re-interpret Ptolemy, to wring yet more truth out of a book that had been wrung dry by the Muslims centuries ago. But the Master brought something new – humility, obscurity. He wasn't physically imposing, not a wonderful orator, not a shining beacon of, well, anything. In the eyes of the world, he was a minor lawyer and failed medic who worked for a morally bankrupt Church in an obscure corner of debatable lands. But the first shall be last, and the last shall be first. Aren't there even more parallels for you to draw here, Most Reverend Philip?

He sheltered me and taught me; this shy, patient, obsessive man with not an ounce of guile in him, such a simple man in the best sense of that word. He didn't have any personal ambition. Can you even imagine that? He didn't care that I was The Enemy – not just an Establishment figure, but from an Establishment he didn't even recognise! He bore with my impatience, my frustration at not being able to understand, my (forgive me) disgust that he had relegated predictive astrology (divinely ordained, as you agree!) to a poor second place after technical astronomy, and even my combative nature. You know that I'm not an easy man to live with, and I can acknowledge that without feeling the need to change myself, but for nearly two years I was his son as well as his student. I learned so much, not just about the Art. I saw what a human being could be, could become. I'm increasingly conscious that I could never live up to him.

I can't decide. I can't decide about anything, Philip. Did I do wrong? He was so reluctant to have his hand forced, for me to spread the word about his wonderful new theories. I pushed him hard, I was relentless, I was charming – if he didn't permit me to write a little book about his ideas, well, I'd go ahead anyway and see what he did! Even the Master turned out to be human. Or was it that I was inhuman? He gave in, as I knew he would, as people always have done to me. Is that charisma that people say I have a gift or a curse? Once the book was published, circulated, read, digested, he simply had to act. He had to finish the work that had been stalled for so many years, before the requests pouring in from every university drove him even crazier than one poor wandering professor from Wittenberg had done!

And he did finish it.

Was that simply because of me? If I hadn't pushed him, would the manuscript of *On the Revolutions* have been lost forever, would history have been different? I'm certain, certain, that the answer to both of those questions is "yes." Things would be so different now, for you, for me, for everyone. I deserve the credit, the blame, the reward, the punishment; but do I deserve his betrayal of me, and am I guilty of the same thing?

You've doubtless heard most of the rest, it's been the talk of Senior Common Rooms and High Tables all across the educated world. He was too ill to see the book through the press, and I should have stayed to supervise. I did stay as long as I could, but I was already late in returning and the letters were getting so urgent, from you, from Leipzig, about the old job and the new job and what was I going to do and when would I be back and what about my students and... and I wanted to stay, but I'm a coward. My life here is comfortable, I get well paid for doing rather little and I didn't want to jeopardise that, even though now I realise that it can't continue, that I can't be an astronomer any more because it's all too painful. I'm no saint – the difference between us, Philip, is that you are, even if you don't believe in the whole idea of saints. And the Master certainly was, and *his* Church still canonises people so there's hope for a St Nicholas of Torun one of these days, once this storm in a teacup has died down.

Which was the worse betrayal? Forcing an old man to complete a book, with my dagger more or less pressed against his frail windpipe, or then abandoning him before the project was complete? Maybe you can set that as a debate for your students. If I had been there, this ridiculous, this thrice-cursed, damned to the deepest Hell, *Preface* would never have been written, never have been included in the book. I am as sure as human knowledge can be sure that it was Osiander who wrote it. It smacks of his theology – "don't believe this book! It's just a theory! Everything's just a theory! The only certain knowledge is that Revealed by God!" It's as bad as anything from those mediaeval Scholastics. It makes me sick, for different reasons than those that would, presumably, appall you (how many different, contradictory, theories does God reveal these days, Philip?) If I had been there... or if the Master had been healthier... if, if, if, it beats in my head all day and night like the drumbeat as I'm marched to the gallows, in my father's footsteps.

So, to his betrayal of me. If it was such. Again, I can't decide and it's driving me mad either way. I'm not particularly proud as professors go, you know that, but I refer you to my earlier point that I'm not a saint. I agonised over the Master's work – understanding it, communicating it, helping him prepare the text for publication. Endless checking of his mathematics. If I hadn't been there, who would have corrected his trigonometry? Not Father Andreas Osiander, that's for sure! The Master was so patient with me, and by the end, so thankful, I was indispensible, I was to be remembered for ever not, perhaps, as a shining star in his new cosmos, but at least as his loyal and predictable satellite. I saw the places in the manuscript where he mentioned me and thanked me and my heart rejoiced as it has never done before, and never will again.

Now I turn the pages of the sullied, doctored, excuse for a book that has set light to the scholastic strongholds of Europe and the stench of ordure from the pages burns my nostrils. Where am I? Where is my name, any of my many names? I have been completely written out of his history, I don't exist anymore. Not even a passing mention of any help received, any encouragement given. I am nothing, and I feel and know that I am nothing. May Christ help me! From the heights of Heaven to the pits of Hell — and all because of a *book*? Philip, believe me when I say that, in all truth, the Master's book means more to me than The Book itself. Peter disowned the Lord, but repented. What if the Lord had disowned Peter?

Was this the work of the same malicious hand at work that wrote the infamous *Preface*? I would love to think so, but my reason cries out against it. Nobody had the authority to alter the text itself but the Master, even I couldn't correct the errors (yes, yes, there were mathematical errors, what do you expect from a lawyer?) without his permission. His mind must have been poisoned by that creeping backstabbing little odious excuse for a theologian – but how could a saint be corrupted by such obvious temptation? My friends – my friend, I should say, or rather young Johan, the only intimate I have here, tells me that the Master was well advised, that he didn't want to drag me into the inevitable controversy, the arguments and passion foreseen by the author of the *Preface*. I would have died for that man! What price the mud thrown on my name by the dogmatists of an alien Church?

Or was it self-protection? The loyal Catholic answered the exam without cribbing from the Evil Protestant? God damn my soul, Philip, which I increasingly suspect to be far from immortal; isn't that even worse? For Heaven's sake, the book is dedicated to the Pope! Which way does that swing things? Yes, he's an astronomer himself, or was before the Cardinals got to him, so he should recognise that truth can be found anywhere – where would we be without Averroës and the other Muslims? But as the Bishop of Rome he has to disown all of us rebels; what have you and Martin done in upsetting so many applecarts that our one holy aim, the discovery of Truth about Creation, is smothered by rolling apples? (Can you at least persuade Luther to put a few apples back, to retract some of the nonsense he's talked about the Master? It must pain you as much as it does me. We both know how prickly he can be, and how much people hang on his words. This isn't the time or the place to develop my argument, but I might not get another chance, so forgive me.)

If I think of the other alternative – that the Master only tolerated me, that all that time he was waiting for me to go away, that I offended him in some way –

No, I can't write about that, not even to you.

I've had my say. Will you receive this letter, Philip? Or will it sit on my desk for a few hours until I burn it, and continue in my cowardly life, living a lie, pretending to care about the Art of Astronomy when my heart is dead inside me? Will I continue here at Leipzig, with poor young innocent Johan, until someone forces my hand like I did that of the Master? Will that someone be you, or one of my rivals here? I'm treading on thin ice here, with Johan, with you, with my conscience, and part of me wants to see what would happen if I crack the ice and fall through, and the rest of me wants to escape as quickly as possible, as far and as fast as I can. I can look after myself, don't worry about that. I could easily learn to be a doctor, a lawyer, even a priest now that morality is valued at such a cheap rate – a new name, a new country and the past is dead to me. Again. How hard can it be to say goodbye to everything that one loves?

You won't hear from me again. Pray for my soul, if that isn't too Roman an idea for you to countenance.

With all best wishes, Georg Joachim "Rheticus" von Porris

Notes on Chapter Thirteen

Rheticus

Fictional characters:

Johan

In fact, Rheticus hung on to his university position for another couple of years after this imaginary letter. Then everything caught up with him, and he was sacked for immoral behaviour (affairs with his – male – students). He disappeared for a while, resurfacing in various places across Europe under various names with assorted professions – but never again writing anything astronomical.

Nearly everything in this account is based on the truth, including the vexed pre-publication history of Copernicus' works. Rheticus certainly knew Philip Melanchthon, Luther's right-hand man and prominent academic. Melanchthon was a keen supporter of astronomy and astrology, writing prefaces to textbooks and actively recruiting the best Protestant scientists for his university at Wittenberg. He was an early exponent of the doctrine of accommodation, whereby a Bible passage could have different meanings for different audiences at different times. (Strict Biblical literalism is a more recent phenomenon. In his letters, St Paul treats Old Testament passages as symbolic rather than literal, and the early Church Fathers continued this trend.)

The most flagrant speculation in the chapter is the content of Copernicus' early draft. We do not know whether Copernicus ever intended to mention his ardent Lutheran disciple in writing, or the reasoning behind his decision not to do so.

14

Isolation

AD 1577

In which we learn that the fabulously wealthy Danish nobleman Tycho Brahe, forbidden by law from marrying his true love, nevertheless lived with her, and established a formidable reputation in his purpose-built astronomical observatory on the tiny island of Hven.

Kirsten Jørgensdatter smiled at the nurse as she crept out of the nursery, pulling the door quietly closed behind her. Little Magdalene was fast asleep, and the nurse had been busy collecting assorted playthings from the floor. Kirsten reflected, as she did most nights, how fortunate she was to have a room several corridors away from them. Dealing with screaming children was not an exciting prospect, nor was an interrupted night's sleep, although for her first eighteen years she had assumed that her fate would be having to deal with precisely those things, as her female ancestors had before her from the dawn of history.

She met her husband on the threshold of their bedroom; he was fighting with a recalcitrant cloak buckle, his thick fingers failing to push the bar home sufficiently, and swearing fluently and profusely.

"Honestly, love, I wish you'd watch your language. Heaven knows how our daughter will grow up!"

She stood on tiptoe to fasten the buckle, and he caught her hands gently, leaning over her to kiss the top of her head.

"She'll grow up fit to work in the world, my dear, able to pass in any company. You're really worried about your own soul, aren't you?"

The pastor's daughter blushed slightly, grinning. He always knew how to read her moods, he always had done. So perceptive for such a lumbering giant of a man, with such a slow voice.

"Have a good night. Anything interesting going on?"

"Not really, but necessary. I'm showing some of the new lads exactly what their duties will be, so I'll be late. Don't wait up. Love you!"

And with another kiss on the top of the head, he was gone, striding along the passageway muttering to himself. Their maid emerged in the doorway in his place.

"Will that be all for tonight, Ma'am?"

"Yes, thank you. See you in the morning."

Kirsten stood aside to let the girl slip past. She watched the neat, precise young woman (what, fourteen?) all the way to her room. She couldn't help but wonder what her own little girl would be doing at that age, and what 'fit to work in the world' could mean for a child so protected by the vast wealth and status of the Brahe family. And then, of course, the inevitable 'what if?' What if anything should happen to Tycho?

She sighed, stepping into the bedroom and closing the door by leaning against its back, the recurring question foremost in her mind; she had thought that she had stopped worrying about it, about what future were they making for their family.

Danish law was very clear on the matter. Since she and Tycho had lived together for three years, any children were legitimate. But since he was a nobleman, and she was a commoner, she could never legally be his wife, and their children could never inherit his titles, status or lands. As every night, she almost started to regret having fallen in love with him; and, as every night, she caught herself just in time.

She threw herself onto the canopied bed, still fully clothed, and lay looking up at the drapery. She did love him, still, so very very much. It could be difficult when he went through his phases of being nocturnal, like an owl, but that had to happen when there were new observers to train. She shouldn't have fallen in love with an astronomer, if she had wanted to see her lover during daylight; that was something her daughter would certainly grow up to know.

Kirsten knew exactly what he would say. "Don't worry about the future, live for the present. Our money will smooth over any little difficulties, and they can't take that away from us."

But that wasn't very helpful, which was why she didn't talk to him about it. He could be so... forceful, so sure of himself all the time, so certain that everything would work out.

She smiled as she remembered their first meeting, in her father's church. He was the new Lord of the Castle, moving back to his ancestral lands of Knutstrop when his own father died, an exotic figure to stride into their small community. A man of the world, rumoured to speak every European language and to be the greatest scholar of the age, and him only twenty-five! His flamboyant clothes, his carefully waxed walrus moustache below his shiny false nose, his huge physical build and booming voice – she had never seen anyone like him before. He had been the talk of the town for weeks before and after his arrival, and soon everybody knew about how he had been raised by his uncle, how his nose had been lost in a duel, how he had studied at all the best universities, how he was setting up an Astronomical Observatory and was going to make Denmark the envy of the academic world, and of course, people being people, it always came back to how hugely, unbelievably, staggering wealthy this young man was, the inheritance of several branches of the Brahe family all flowing to him.

She rolled that name around in her mouth, murmuring it aloud. She was always amused when new staff started work here, from the German lands, from Poland, from Scotland and England, from France, all struggling with their Danish and pronouncing his name in their horrible Latinised way, "Tie-co Bra-hay." They only ever did that once, before he would slap them on their back hard enough to send the scrawnier ones flying, bellowing "Tee-cuh Bra, man, you're in Denmark now!"

How many astronomers did he have working for him now? It must be thirty at least, she lost count of the comings and goings.

She should be proud of all that he has done, she told herself. And yes, she *was* proud, of him, of this place, this Fortress of the Heavens. The best minds of all Europe were coming here, to Uraniborg, set on this tiny little island in the middle of nowhere, away from lights and distractions. And away from female company, and polite society, and anything for young children to do. They ate and drank stars and planets, along with all that ale and wine; it was a wonder they could see straight along those quadrants and sextants and octants and astrolabes and Jacob's staves and backstaves and... and everything that Tycho had designed so painstakingly. Where did she fit into it all, and their little girl?

Reaching up behind her, she pulled down one of the feather pillows. She ran the household, she welcomed the guests, she kept the day-to-day things ticking over, she told Tycho when he'd gone too far, pushing people beyond their limits. And she listened when he talked about his new theories, although he never went into the details with her, and she didn't know whether to be relieved or offended by that. And she was always there for him when he wanted to rant about how stupid everyone else was being, and to cuddle up to him when they did get time alone together, and to soothe him, and to make love, and to remind him of the bond that they shared. Yes, she did love him, and she was proud and pleased that he chose her instead one of those stuck-up young noblewomen who were chasing his fortune and couldn't appreciate his genius or tolerate his eccentricities, and she knew that he loved her. Think what he had sacrificed! The family name might die out and the estates be split up, because he loved a poor preacher's daughter. But that didn't mean she found it easy.

It had all been so different when they started out together. Even her father had approved when Tycho swore to him that he would be faithful to her. He took her with him to Court, and nobody said anything about their... irregular life. She had heard that the King himself wished to live with a commoner until his Council talked him out of it, so the courtiers understood this sort of arrangement very well. Then there had been the townhouse in Copenhagen, the glittering banquets, the social whirl – and their tour of Europe, Tycho meeting his fellow astronomers and little Kirsten here, the little Kirsten who grew up in plain linen smocks, had chosen dresses in the latest fashion from all the latest designers, with money no object, and the arrival of the children, one after another, and the tears when they buried two infants – she had hardly had time to stop and think, so no wonder she felt confused. No sooner had she learned to live one life than another approached, and then... then the move to Uraniborg. Not at all what she had expected from any of the lives she'd lived in the previous five years.

She thought back to the arrival of The Letter, from Frederik II himself. It had taken weeks for it to catch up with their progress from court to court, university to university, on a tour which had no definite project, itinerary or planned ending. Yes, although she had enjoyed seeing the cities she would otherwise never have seen, she had pined for home, for her surviving child (left safely with the nurse), for some sense of rootedness such as she had known in the happy confusion and noise of the Preacher's House. The Letter had suddenly promised that.

Tycho had been declaring, to anyone who would listen, or failing that to anyone within earshot, that the state of astronomy was a complete mess. Some argued for Ptolemy, some for Copernicus, some for their own systems.

"What will decide the matter," he would roar, smashing his tankard down on the table, "are *facts*."

Every great astronomer of every age had been a mathematician at heart, he would go on to explain (she'd heard the speech so many dozens of times she had it by memory), who took a mere handful of observations of the actual positions of the stars and planets themselves. The lists of movements, of positions, so carefully compiled over hundreds, thousands of years, which these armchair astronomers relied on – all of them were useless, utter rubbish, a mere waste of papyrus and paper and ink, since they were all so inaccurate, measured with equipment that could hardly be expected to produce good results. The theorists were building intellectual castles in the air, using 'facts' from cloud-cuckoo-land – what was needed was a real, solid, physical castle, an observatory big enough to hold the most up-to-date, state-of-the-art instruments (which he himself would design) to measure the positions of heavenly bodies accurately, and a whole team of dedicated observers who would systematically record everything about the night sky, every night, for a few decades. Then, and only then, could the mathematicians and theorists have any right at all to talk about the heavens. Then, and only then, could the truth shine through from the mass of lies, shadows, and misunderstanding which had cloaked it from the beginning of time.

Kirsten had never been able to work out whether this was all bluster, or a real proposal for a project, or whether Tycho ever distinguished between the two in his mind. (You might say the same thing about living with a commoner simply for love!) Who, after all, would be mad enough to build a castle, far from any city, big enough to house a hundred people, all for the sake of astronomy? Even the Brahe family fortune didn't approach the levels needed for that!

Then came The Letter. Frederik was mad enough to provide the funds, and the lands, for the glory of Denmark (and his own reputation as a patron of scholars) – and suddenly Tycho added Lord of Hven to his other titles, and was absolute lord and master of this tiny chunk of rock with its pitiful fishing village and handful of struggling smallholdings, with a free hand to design and build as he chose. A settled, rooted life at last! On a rock. With a whole constellation of shy young astronomers, just out of university, who barely spoke Danish and blushed whenever she approached.

At least it was a very comfortable rock. She snuggled down against the pillows, idly unlacing her simple workaday tunic as the familiar thoughts rolled around and around in her head. They dined well, with banquets twice a week: they had all the servants they could have wished for, and Tycho was happy in his work. He was thinking of so many new theories, but was always waiting for more measurements. Her poor darling, he didn't want to publish until the theories were perfect, but also didn't want anyone else to get there before him. He was already talking of extending the building works, and the third floor of the west wing wasn't even finished yet! Sometimes she wished he would just slow down. There were to be the Medicinal Herb gardens, and a new Observation Wing, and the Great Wall, and improvement work on the farms, and now he had started talking about not only a printing press but also a paper mill to feed it, and that would need more accommodation, and more servants... couldn't they do one thing at a time? He doesn't understand her worries – "Strike while the iron is hot, my girl!" And she had known he was like that when they met; their courtship had only lasted a month before somehow she found herself under his roof, sharing his bed.

The months that had followed The Letter had been a particular low-point. Living in the tiny inn in Hven village, Tycho had been out in all weathers obsessively pacing the site, taking measurements, ensuring that the castle was aligned precisely to the points of the compass. 'Obsessive' was a good word for him, in every respect. How could a man so swayed by emotion, so prone to sudden action, also be such a meticulous planner, observer, builder? She would never understand him, and she knew that she would never stop trying. Everything around him must have seemed to move so slowly compared to his racing mind, and he could vent that frustration to her – perhaps above all else, she consoled herself, that's something he could never do if he didn't have someone to love, and to love him.

Kirsten rolled slowly off the bed, undressing and putting on her long nightgown. She left her clothes neatly on a side-table, unwilling to make extra work for the maid-servant whom she could so easily have become herself, had things turned out differently. She was tired, too tired to brush out her hair or otherwise undertake her normal routines that would help to calm her before sleep. She'd been this exhausted for a week or so now, and she wondered whether another child could be on the way. At least she and Tycho agreed that a large family was desirable and a blessing to the parents. Perhaps she would get to see this one grow up, not like poor Kirstine and Claus, gone before they had really arrived.

What did they disagree on, then? Nothing, in the essentials. They both knew the value of work, of using their God-given talents however unusual they might be. They were prepared to do what they believed to be right, even if it was uncomfortable. Both of them loved society, but also solitude. Yes, even the flamboyant clothes, the desire to be noticed, to be talked about – she knew that was a part of her. They were both good Lutherans, who valued education, and they were both loyal subjects of a king who had been more than kind to them, in a country they could be proud of, which had avoided the excesses of many of its neighbours. They laughed at the same ridiculous things, they hated pretence and play-acting, they knew the importance of being true to themselves. They loved each other, and Kirsten kept returning to that because it was the very anchor of her soul, the thing that stopped her from being carried away by the strength of her thoughts. They loved their child, and wanted the best for her. So why did she worry?

Tycho was strong, full of self-belief. Or was he stubborn, full of self-importance? He didn't care what others thought – he had argued with everyone, including those long-dead like old Aristotle and Ptolemy, and he was offhand even with the king, his patron. Was she worried because she knew that, deep down, he didn't care what *she* thought? Maybe that was it. He rejected people with a wave of his hand and a laugh, he had no time for opinions that weren't backed by extensive observations and analysis. What could she offer him, other than physical comfort, a shoulder to cry on and an ear always ready to listen? What did he really think of her? Was she simply part of his entourage, albeit an important one?

No, she couldn't think like that. He had sacrificed hundreds of years of family history for her, put his own good name and judgment at risk. Why would he have done that, if she had been no more than his mistress? Good Lord above, she needed to find some way of breaking this cycle of thought. Was she happy here?

She moved back to her bed, rolling herself in the full width of the sheet. Tycho wouldn't be coming to bed until daylight, given how thorough his talks to the new astronomers were, and she really couldn't wait up – she had to be ready to welcome a bright young thing fresh from Oxford, or Cambridge, or some such far-off place. Maybe one day they would visit England, or perhaps Scotland was more likely? He had told her that their king was very fond of scholarship, and might even be persuaded to visit them on Hven. Imagine that, Kirsten Jørgensdatter playing hostess to a king! She would have to order a whole new wardrobe from Copenhagen for the event, if it ever happened, for herself and for Tycho. When they had spun tales about their futures, as girls back in Knutstrop, nobody had ever dared to imagine that.

So, was she happy? Yes, she was. The Lord knew, she was grateful for everything that he had done for her, for the comforts, the experiences, the love.

She was starting to sink into sleep, her lids closing as she folded her arms across her chest and turned her head to one side, cheek sinking into the cold, welcoming linen of the feather pillow. Yes, she was happy. But did it all have to be on such a barren rock?

Notes on Chapter Fourteen

Tycho

Fictional characters:

None

Tycho Brahe must be the most colourful character in the history of astronomy. He is normally referred to simply as Tycho and his theories are referred to as Tychonic, rather than Brahean or anything similar. The biographical information in the chapter is all true, remarkable though it might sound.

His contribution to astronomy and cosmology is often seen by historians simply as collecting huge amounts of accurate data (in itself a novelty). However, it extended far beyond that by changing the way that people *thought* about the subject. Tycho realised that observations and theories lead to each other in a virtuous circle. There is much to support the claim that his observatory of Uraniborg was the first modern scientific research institute in any recognisable sense, dedicated to all aspects of astronomical endeavour.

Moreover, Tycho's undoubted arrogance was in marked contrast to the timid attitude of many university professors. He rarely qualified his statements, and was prepared simply to assert that his great predecessors were mistaken. He saw no reason why respect should be given to a writer simply because he wrote a thousand years ago, and thereby prepared a path for the wholesale reform of astronomy over the next couple of generations.

15

Parting

AD 1597

In which the story of the Brahe family continues, with a reversal of fortunes as unexpected as it was complete, and light is shed once more upon the parallels between the rules governing the motions of the planets, and those governing the motions of our hearts.

The hall looked so deserted, almost forlorn.

How different from when we arrived, a month ago.

Anders glanced around the building, trying to remember where everything had been. Dust was already beginning to settle on the areas where the printing presses had stood.

Strange how much larger the room looks now. It's been an interesting posting.

The young soldier sighed.

"I think it's clear, now. Want to come in and check?"

"No thanks, I'll wait here. It's all too depressing."

I think I've caught some of your depression, my darling. This is just a job for me, but it was your home. You grew up to the noise of the presses, the smell of the ink, the chatter of scholars. What a life for a young girl!

"Fine. I'll just be a moment."

He stepped through into the compositor's office, mechanically scanning the room to see if there was anything left. He knew that there wasn't; he had worked hard at dismantling and lifting and packing until his arms ached and his back was sore, but he had told the sergeant that there was still a morning's work to be done. One last, precious, morning with his Sophie.

The detachment would be returning to the mainland tomorrow, duty done, back to his life of barracks and square-bashing. Sarge had said that His Majesty planned to demolish Uraniborg completely, to remove every trace of the astronomical fortress, but that was a mission for engineers, not ordinary footsoldiers. Escorting the family to the border and seeing them off – that was a mission for Palace Guards, not ordinary footsoldiers. Ordinary footsoldiers were good at dismantling, lifting, and packing but very little else, it seemed.

Not that I'm complaining – I've had a month with a beautiful girl. Is it coincidence that she's wearing that dress today, the sleeveless cream-coloured one that I praised so highly when we first met, stumbling against each other on the steps, my arms full of books from her father's study? They tumbled and fell, we picked them up together, and all I could think of to say was that she was wearing a nice dress. Smooth, Anders, smooth. I'll bet she's forgotten all about that. But it must have made an impression somehow because, well, here we are together.

"Come on, hon, I don't want to be alone out here."

Anders nodded to himself and pulled down his liveried tunic sharply. It had been hot work, and the squad were in undress uniform. A shame, since he knew for a proven fact that girls admired him when in full ceremonial rig.

"Coming!"

He pushed the door closed behind him, shaking his head. Then Sophie's arms were around his waist, and his gloomy thoughts disappeared like night-terrors at dawn.

They lay sprawled alongside each other under the summer sun, fingers interlaced. Anders turned his head to gaze at her long, rich, dark hair, spread out like a saint's glory in an old painting, and her alabaster skin, the complexion of one who mainly stayed indoors.

Her eyes are so deep, so clear... eyes that are far too wise for someone so young.

"What are you looking at?"

Her voice was barely a murmur, low and smooth like that of her mother.

"You, my darling. I want to fix your image in my mind."

"That's sweet. Shall I lie to you, say that I'll write, we'll keep in touch?"

He shook his head, grinning.

"It's been a wonderful month, Sophie, but I can't see you taking up with someone like me. We both know that."

A squeeze of his hand.

"We're both commoners, you know. That isn't an issue between us."

"Hah, very funny. Some of us are commoner than others. Your big sis, Magdalene – she's engaged to a medical man, not to a footsoldier. What would your old man say?"

"You'd be surprised. Mum isn't from a good family, you know that."

She rolled onto her side and kissed his forehead.

"Shall I play the romantic heroine, tell you that I'll be lonely, that I'll ache from my broken heart?"

"No, don't do that. With all those brothers and sisters around, I'd expect loneliness would be something new and exciting for you. Besides, there's your dad's whole retinue."

"Bah. Servants, or astronomers – I don't know which are worse."

"The servants aren't so bad. Old Jepp is a pretty classy act."

Tycho had sent his dancing dwarf jester to entertain the soldiers, to show that he bore them no personal grudge, that he realised they were only obeying orders.

In fact, the old man has been more than fair, given that we're here to dismantle his life's work.

Anders thought back to the briefing that Sarge had given them, the day before they embarked on the short ferry crossing.

"Old Man Tycho is rich and powerful, living it up on his island like a king from Once Upon A Time. They say he's brilliant, but he's fallen out with the real King, and this isn't a fairytale. So we're chucking him out of his little kingdom, and our big one, and we're going to be the removal men. Nothing gets hurt in the packing, though, understand?"

Nothing gets hurt, sure. What about the people, though?

It hadn't taken more than a couple of evenings with the locals at the inn to establish the story. They all knew everything that went on at Uraniborg – the long, boozy feasts, the late-night astronomy, the stand-up fights about mathematics, the comings and goings, the pet elk (sadly now dead after an accident involving beer), and most of all what Tycho, their Lord of some twenty years, said and did. It always came back to that. Tycho was some kind of superman to them, putting their little scrap of land on the map, bringing fame and fortune even if he *did* charge them more rent than the previous landlord. And encourage them to learn strange trades such as papermaking. And… but back to the point, after some more beer had been purchased. Tycho had been Frederik's blue-eyed boy, could do no wrong. Then the king died, as kings do, and the new king didn't like Tycho, and Tycho didn't like the new king. Christian IV expected Tycho to behave seriously and to do the things he was paid for – turning up at court to advise His Majesty on philosophical matters, casting horoscopes, writing almanacs, looking after his fief, looking after the chapel whose revenues went to his pocket… in short, to be a good courtier, a good nobleman. Tycho didn't give a fig for any of that, and told young King Christian as much, in so many words. The reaction was inevitable.

Now little Hven would lapse back into obscurity. Did you know that King James of Scotland had visited them?

"A penny for your thoughts."

"I was just thinking about your dad. He's done such a lot for this place. The fortress... breathtaking work. It seems like such a waste."

Such a waste. The building was magnificent, and so richly decorated. The sheer scale of the project was almost beyond belief. Astronomical instruments built into the fabric of the palace: a whole wall turned into a quadrant, with a beautiful mural of a much younger Tycho in the middle. The old man certainly wasn't modest, his portraits were everywhere. But that's easy enough to forgive, since there was a gorgeous painting of Sophie in the entrance chamber... or had been, until he had taken it down to pack.

"Soph, that portrait of you that used to hang in the lobby. How old were you?"

"It must have been done... three years ago? So just sixteen. You liked it?"

"The painter couldn't capture your beauty. I'm glad you wear your hair long, now."

She grinned again at him, rolling again onto her back and staring at the cloudless sky.

"One day you'll tell your children that you kissed the gorgeous Sophie Tygesdatter, the one that married the Emperor of China and lived in a castle made of finest porcelain. What will you tell them about me?"

"That you were unfairly banished by a cruel and jealous King, who could not stand your ravishing good looks. Does that sound good to you?"

"I don't like the sound of 'banished.' Can you say 'exiled' instead? It sounds much more romantic. Anyway, is it an exile? Dad is always going on about his student days, his Grand Tour soon after he met Mum – what have I seen? Hven, and Copenhagen. I'd love to see Paris, Petersburg, Prague… maybe not Prague, come to that."

"And what's wrong with Prague?"

"Dad's got enemies there. Some bastard stole his theories, he's passing them off as his own, they're fighting about it. Pity, it's meant to be such a beautiful place. Have you ever seen it?"

Anders laughed.

"Me? I'm twenty-one and a footsoldier. How would I see Prague?"

"I don't know. For all I know, we could be at war with the Holy Roman Empire. You could have fought a heroic battle there, defeating the Forces of Darkness, or something. Dad never talks about politics, he says it's all pointless posturing. The sort of pointless posturing that could get a fine, brave, and oh-so-handsome young solider like you killed."

She squeezed his hand again.

"My brothers want to join the army. Dad's so disappointed, but I bet his folks weren't pleased with him when he took up astronomy. Do you lose your memory when you become a parent, do you suppose?"

"They're young, they'll grow out of it. Some of us had no choice; your family will never starve, wherever you end up."

She turned to him, shifting to snuggle alongside, her arm flung over his chest.

"Maybe I should protest, but you're right. Dad will find another patron, some prince or king or emperor, and we'll be back in business, just in another language. Don't worry about me."

He placed his hand over hers.

"Oh, I won't. So what makes your dad so valuable?"

She laughed easily.

"He's shown that everyone before him was wrong about the

heavens, that's all. Didn't you know that he's brilliant? No, really, he genuinely is. Behind all the bluster and show, he's the greatest mind of our time, and I'm not just saying that because I'm his daughter."

"Mmm. Tell me."

"Well, you know that the planets are attached to crystalline spheres?"

"They are?"

"Heh. No, they aren't. But everyone for the last two thousand years thought that they were."

Anders sat up slightly, leaning on his arm. He ran his fingers over Sophie's hand. "And...?"

Sophie also sat up, shaking out her long hair.

Mmm... I'm going to miss those beautiful locks, the way they feel against my face. I'll have to find myself a girl when I get back to barracks, although she won't match up.

"And so everyone for the last two thousand years was wrong. Doesn't that impress you?"

Anders' turn for the laugh, the smile.

Why am I talking about astronomy on our last morning together? But it has been her life, her world, since she was born. Relax, Anders, go with it. Maybe you will stay in touch, after all, maybe something could happen between us to keep us together...

"Sure, of course. How does this place tie into all that?"

"Observations, silly, relentless observations. All the kit you dismantled, it's not just for show. Dad and his team measured everything. Some of them were sweet, his helpers, but none of them were... like you."

Whatever that is supposed to mean.

"Do you love me, Soph?"

Shit, had he said that aloud?

She was tracing an intricate pattern on his tunic with her free hand, pausing for a long time before she spoke. When she did, her voice was calm, low, level, patient.

"So, for two thousand years, the planets are stuck to circles, stuck to spheres. The heavens are unchanging, eternal. Then Dad sees a new light in the sky, no big deal, people do from time to time. Everyone says it's just stuff in the air, fire in the air somehow. Except this one isn't, it's too far away. Dad measures it, finds it's too far away, so far away that it must be a star. A new star! In the unchanging heavens. Do you follow?"

"Yeah. Sophie, listen, do you love me?"

"Shush, I'm trying to answer you. In my own way. Everybody thought they understood about the difference between Earth and heaven, one can change and the other can't. Wrong! Everything's the same. Then there's a comet. You know what a comet is?"

"Yes, of course, but Soph, darling, I don't..."

"A comet, tracing its path through the heavens. Dad works out what its path was. It goes past all the planets, up to the Sun, and then away again. Do you see? Do you see what that means? It comes from the stars, goes past the planets, then goes away again!"

"I... no. Look, Soph, I can't..."

She sat upright, flicking her hair back with both hands.

God, she's lovely. Can't this moment last?

"It goes straight through the spheres, the circles, the whole damn mechanism. Like they're not there. Because they aren't! Dad has shown that they aren't there, that the planets wander freely! Sure, they follow some rules, something steers them in their courses, maybe angels, who knows? But they aren't fixed to anything."

Her eyes were shining, she was staring at him imploringly.

"Don't you *see*?"

Is this some kind of test? If so, I think I'm going to fail. What has any of this got to do with us, with people, with love? Is it better to be a liar or a fool? No lies. I owe her more than that.

He shook his head, slowly, sadly, eyes darting guiltily to the side, then down to her body.

"Oh, Anders. We're like planets, wandering through empty space, all alone with no foundations. We're steered by the incomprehensible rules followed by our souls, our minds, our angels. But there's nothing that forces us to move in certain ways, no machinery that sweeps us along. Of course I love you, silly boy. I'm not going to ask if you love me, because I don't think you know. Now, kiss me."

"Do you think I should say anything to her?"

Kirsten was looking out of the window at the passing countryside. It did seem very bleak compared to the well-ordered farmlands of Denmark. Frankly, she was worried. She had never seen Sophie like this before. They had only been on the road for a couple of days, the long procession of carriages, wagons, horses, pack-mules heading God-knows-where, and the girl had been in floods of silent tears for most of the time. The younger ones had certainly been unhappy to leave, but were generally taking their cue from their father, facing the road ahead as an adventure.

"No, my dear, I think you'd best leave well alone."

"I just didn't know she felt so strongly about the place. Whenever we visited town, she wanted to stay on, talked about her Island Prison..."

Tycho snorted, and put aside the papers he had been trying to read, as much as the jolts of the carriage on the poor roads allowed.

"Not the place, angel, but the people. A person. That young red-headed soldier. Didn't you notice?"

Kirsten blushed. She had always prided herself on reading people well, but she was getting old. They were all getting old. And the children... Sophie was nearly the age that she had been when she first met Tycho, but she still thought of her as a little girl. How foolish!

"No, I... young love. What woes it leads to! Who would ever wish to feel its pangs and its pains?"

She rolled her eyes melodramatically, throwing up her hands.

"You're right. I still haven't got over it, and I think you're afflicted too! Anyway, I trust Sophie. She's got a good head on her shoulders, she's bright and she's pretty, and I'm not going to press her for any details. If she's got herself into trouble, she'll tell us if she wants to, or needs to. She'll learn from this and move on. It wouldn't work, anything steady between her and a soldier."

"Oh, listen to Lord High-and-Mighty... in exile, I might add, of No Fixed Abode."

Tycho laughed, his face wrinkling. He stroked his moustache, now as white as any walrus-tusk.

"No, angel, I didn't mean it like that. The centre of his orbit must be Duty, while we all circle Truth. Just as the planets orbit the Sun, and it circles around the Earth; paths may cross, but are then swept away. And Sophie? Sophie's a comet, burning bright, her hair streaming out as she moves on paths we can't calculate, slicing through the shattered remains of the spheres. You know, I ought to ask Chris to piece together the mathematics of people, as well as of planets. It's all the same thing, a point I've made to Sophie many a time. I wonder if she listens."

"Do you ever wish she was more like Magdalene, settled, steady, predictable? Or like Lisbeth, calm, reliable, accepting?"

"Heavens, no! We've got seven kids, let 'em be different. Could you pass me that booklet by your left elbow? The one with the calf-leather half-binding? Thanks, angel."

And further down the convoy, Sophie slept fitfully, dreaming of the days when wanderers were fixed to solid, heavy, reassuring spheres, if only you had the wit to see them.

Notes on Chapter Fifteen

Tycho

Fictional characters:

Anders

Tycho's later life was as complicated and extraordinary as his initial rise to prominence. After a happy period on the island of Hven, with an ever-growing family and circle of astronomical colleagues, his neglect of his duties and his prickly nature finally led to his forcible eviction, as described. The romance of this chapter is wholly fictional, however.

In his Uraniborg years, Tycho and his team gathered more data than had ever been collected by European astronomers, to unprecedented levels of accuracy (remembering that this was before the invention of the telescope). This consisted of series of observations of planetary positions, and also the most precise measurements of the positions of stars yet recorded.

Side-projects involved the study of medicinal herbs, alchemy and large-scale astrological researches (attempts to correlate planetary positions with weather conditions). His main theoretical contributions were the shocking realisations that the path of a comet takes it through inter-planetary space, wrecking the theory of impermeable spheres once and for all, and that novae, rare bright lights in the sky, were stars rather than atmospheric effects (as Aristotle had believed), undermining the theory that the heavens were perfect and unchanging.

Tycho developed his own compromise system of cosmology, with the planets orbiting the Sun but the Sun orbiting a fixed Earth.

This allowed him to use Copernican mathematics (neater than Ptolemaic), without adopting the whole system.

16

Collision

AD 1600

In which Tycho crosses swords with Johannes Kepler, a talented, shy, and devout young mathematician looking for work, and we find that sometimes the sparks from a head-on clash can be precisely what are needed.

February 4th

This is ridiculous. I'm a grown man with a family and a good job, and I feel just like a schoolboy waiting to see the headmaster. In a sense, that's what I am... he holds all the cards, he's everything I want to be. Rich, famous, and so very brilliant.

John Kepler tapped his foot nervously on the stone flagging.

Interesting echo... where is that secondary coming from? Surely the ceiling is too high..? No, man, focus.

The heavy oak door swung open, and Tycho was standing there. Kepler had seen his picture before, of course, but in the flesh he was larger than he had supposed, a huge presence, so richly dressed in velvets and furs.

"Kepler? Come on in."

His voice was deep, growly. From all that he had heard, John had expected much more... well, gentility. Nobility. Manners, at least.

I've waited so long for this moment. He liked my book, we've exchanged some interesting and courteous letters, he invited me here – so why do I feel so nervous?

The younger man stood up from the cold stone bench. He shivered, pulling his travel-worn cloak around him.

I really should have bought some new clothes after a month on the road, but that would have meant another day's delay.

Without further ado, Tycho turned on his heel, disappearing back through the doorway. Kepler glanced around the hallway, unable to shake the feeling that he was being watched, and followed his potential mentor and employer through into a small study, cluttered with books and paper. Tycho sat heavily, reaching over to a crudely-printed pamphlet and pulling it towards him.

"You didn't get my letter from two weeks ago?"

"No, sir. I was already on my way. Any letters would probably be following me from inn to inn."

He shrugged apologetically, hands spread.

"You know how it is."

"Yes, I know how it is. It means you can explain in person, rather than by letter."

His voice was cold, distant, his eyes boring into Kepler's skull. There was a sense of barely-controlled menace in the air.

"Umm… explain? I don't…"

"Explain *this*."

Tycho pushed the pamphlet, open at a particular page, to where Kepler stood. His eyes did not leave the young hopeful, watching for his every reaction.

Kepler looked at the document, poor quality ink smudging on the paper. There was a passage in specially bold print:

"And I can claim as one of my supporters in this fight, JOHANN REPLER, Provincial Mathematician of Graz, who has written to assure me of his devotion and to confirm the accuracy of my observations and calculation."

Obviously him, despite the misprint. Puzzled, he flipped back to the title-page. *A booklet by Ursus?* The former Imperial Mathematician, Tycho's predecessor as adviser to Rudolf II, whose protracted fight with Tycho had become the favourite entertainment of all the astronomers of Europe, as the insults got more and more personal.

Why would Ursus be mentioning... oh, dear Lord. No, no, it couldn't be...

The bottom fell out of Kepler's world, his vision swimming as he dropped the book and clutched the table.

"I... I can explain..."

"No, you can't. Get out."

Perhaps it would have been more tolerable if Tycho had shouted rather than whispered. John was a schoolteacher himself, as part of his duties, and always found the calm, quiet word far more effective than the bellow. Just like a schoolboy before the headmaster, and the headmaster wasn't interested in excuses, even if they were true, only in administering the punishment and moving on.

Kepler was normally pale, but now he was deathly white.

Am I going to faint? Can I at least leave with some dignity intact?

He nodded once, gathered his cloak around him, and walked quietly to the door, faint stars at the edge of his world. There was a pounding rush of blood, a sinking feeling in his stomach. At least he managed to get outside and push the door closed, so that it was one of Tycho's aides who found him slumped on the floor rather than the great man himself.

In the cold, anonymous room of the cheap inn, he sat down to write while the sounds of drunken merriment filtered up through the floorboards.

To Lord Tycho Brahe, at Benatky Castle,

I do not attempt to excuse myself and will not attempt to lie. I wrote to Ursus, offering him my service as a mathematical assistant. However, you must understand the context.

I wrote to Ursus almost exactly the same letter that I wrote to the Landgrave of Hesse-Kassel, and to my former teacher Mästlin, and to Clavius – and to you. I have read all of your books (yes, and those of Ursus and the others) that I could find, and I wished to work with any one of you, to be an assistant to someone whose name was famous. I was young and impetuous, and having written my *Mysterium* thought that I had proved myself in the eyes of the astronomical community, that I was ready to spread my wings and fly higher than teaching rich brats and drawing up horoscopes for fat councilmen. So I wrote to all of you, and sent all of you my work. You will remember that this was years ago now. You, my Lord Brahe, were the only one of these great (and not so great) names in astronomy to give me encouragement, to take this wretchedly poor, provincial, scholarship boy seriously. Ursus never even replied to my letter. How could I know that he was storing it up as ammunition to use against you? I was so naïve back then that I did not even realise that the two of you were at loggerheads, that he (as the world knows!) stole ideas and even papers from you, treacherously, when you had offered him hospitality in your famous observatory in Denmark.

If you decide that you never wish to see me again, then I will honour that decision with sadness in my heart, and head back to war-torn Graz and my family. But if you can find it in your soul to exercise – no, not forgiveness, but a chance, just a chance that I might redeem myself in your eyes, I shall remain at this address for the following three days. I have many books to read, acquired on my travels here, and will await any reply with trepidation and eagerness mixed in equal measures.

Humble duty, etc, etc.

No, that didn't hit quite the right tone. Throw it away. Time for the third attempt. Why won't they shut up downstairs?

February 22nd

Tycho,

Longomontanus and Tengnagel have been running over the theoretical, mathematical and observational aspects of your astronomical investigations with me, these last two weeks. I am sorry that I have been more active during daylight hours than you – I wish with all my heart that I could take readings along with the rest of your team, but you are aware that my frail constitution does not easily tolerate the night air – I nearly died several times as a youth, which perhaps lent me an unusual perspective on the universe – and that my eyesight is weak. I assure you that, if you do decide to allow me to become a permanent member of your staff, my mathematical vision is unclouded. Longomontanus was rather unhappy with me when I pointed out some of his errors, but the Truth must come first, now and always. So, thank you for allowing me to take up so much of their time with my questions.

I have been considering in detail your arguments in favour of maintaining the Earth at the centre of the universe and rejecting the theories of Copernicus. I certainly agree that the poor cleric based his constructions upon observations which are hugely inaccurate compared to those which you have amassed, but I do not think that is the root of the problem. Indeed, I am working on a demonstration that the mathematical predictions of planetary positions given by the Ptolemaic, the Copernican and the Tychonic systems are exactly identical, corresponding solely to the choice of viewpoint of the observer. One could, if one so wished, work out the movements of the planets as observed from the Moon, and doubtless any inhabitants of that arid sphere would consider themselves to be the centre of the world. No, no, my lord – the arguments for and against the Earth's motion must be either physical or theological, not mathematical.

You rest much of your case on the lack of observed stellar parallax. If the Earth were moving compared to the fixed stars, you argue, then the stars would seem to move relative to each other. As I walk along a path, the apparent angle between two distant mountains will change. If I walk far enough, two mountains that started side-by-side might even align. Yes, so far this is true. But what if the sphere of the stars were immeasurably distant, the mountains so far away that we detect no change as we stroll over the plains? Cusanus suggested something like this many years ago, and I have read no convincing counterblasts to him based on physics, although there are possible theological objections.

This brings me to a more important point of disagreement between us. Your theology, frankly, confuses me. Can we meet to discuss your views? I would be more than willing to stay up late after dinner and talk before your first observational shift, on any day of your choosing.

Best wishes, J. K.

April 12th

Tycho,

I have taken a room at the Golden Fleece inn, just off the Stone Bridge. Prague is a beautiful city – far more secure than Graz, and a lovely place to raise a family – and I have nothing but admiration for the work you are doing for the Emperor out at Benatky. However, under the circumstances I do not believe that I can work with you. I intend to remain here for one further week before the long journey home: if you should change your mind about my potential terms of employment, you know where to find me.

You were shouting too much at our last meeting to hear me calmly, so I am setting forth my requests in writing. If I know you at all, you will scan this letter when you receive it, toss it to one side, and reply at the last possible minute. It has been an extraordinary revelation to me how a man who is so meticulous about his astronomy can be so disorganised when it comes to everyday matters, and to common courtesy. You might have been born with a whole drawer of silver cutlery in your mouth, my lord, but we are all of us humans and entitled to respect regardless of our opinions.

Obviously, the main sticking point is money. To put it bluntly, I am a stronger mathematician than Christian Longomontanus. Although he's a nice enough person, and ten years my senior, he doesn't have a wife and child to support. Neither does he have to relocate from halfway across Europe. Your 'full and final offer' of a salary which amounts to two-thirds of his is, frankly, an insult. I would not consider working with you until and unless my salary were to be the same as his. How would it look for your assistants to be shabbily dressed and half starved? The money that you have spent on merely one of your precious armillary spheres, or your ridiculous stellar globe, would pay what I request for a couple of years. You probably spend as much on buying wine and ale as you do on your whole staff, which reveals an interesting set of priorities. They are happy to work with you, whereas I... I could easily find an astronomical job in any major court in Europe, now that I am determined to ascend to better things. The question is, do you want me on your team or not?

Secondly, there is the vexed question of access to the observations you have made. You were kind enough to recognise that my *Mysterium Cosmographicum* contains many interesting ideas, which could only really be verified by checking against the storehouse of data you have accumulated over the years. And yet you deny me the chance to look at the measurements! Everything I require has to be written out in triplicate, and sent to Longomontanus and Tengnagel for them to check in your Holy Books, concealed in a strongroom I know not where. If my pen has slipped and I have written down an incorrect date, I lose four whole days of work while I repeat the process. This is ridiculous. I know that you were burned by the whole Ursus affair, and are jealous of your own honour and discoveries, but astronomy is not – or should not be – a Mystery Cult, where only the initiated priests can be told the Truth. It is – and you fail to understand this – the study of the nature of God as it is revealed in the structure of the physical world, and by your restrictions you are preventing others from discovering this Truth. I believe that the Bible (an interesting book, you should perhaps read it sometime) has something to say on this matter.

Thirdly, your dissolute lifestyle tends to bring your whole enterprise into disrepute, and by association tarnishes the reputation of your staff. You should hear what they say about you, here in the city! You're "Old Noseless", the drunken astronomer who pops up to tell the Emperor about his horoscope and then disappears to continue drinking in his castle. When did personality become more important to you than philosophy, or have you always been like this?

Fourthly, your light-hearted attitude to religion, and to religious authority, alarms me considerably. Like it or not, this is a world in which people are dying because of their faith, and Europe looks likely to become embroiled in a general conflict between the Catholic and Protestant states. In this political climate, it seems extraordinary to me that you treat the whole notion of religious devotion as something like a bad joke. You will say that my views are coloured by my own personal experiences, and of course they are – but so are yours, my fine friend. From a matter of personal safety, Prague is a very good place to be under the present régime. I am a member of no Church, since I firmly believe that Luther and Rome are both in error, in different ways. This puts me in a difficult position as both sides (why do they have to be sides?) treat me as an enemy. To be laughed at by you, who seem to have no recognisable faith but are happy to call yourself a Lutheran, and to be called a "weak-minded fence-sitter" is demeaning and insulting. Regardless of any other arrangements we might be able to make, I formally demand a written apology for your insults.

Fifthly, your insistence that the Earth must be fixed in place is in defiance of all physical sense. However, this is a point which I cannot argue effectively before I have completed my new study of the motion of Mars, and that is something which I simply cannot do unless I have money enough to live, unfettered access to your observational data, respect for you and your work, and an apology from you. I know that you are curious about what I am doing. Are you prepared to back up your curiosity with money, a small amount of self-abasement, and a promise to be more reasonable and calm?

I hope that I hear from you again. If not, I shall visit you before I leave to collect that written apology, but I have no particular desire to speak to you.

In sorrow and anger, J. K.

April 14th

John Kepler removed his brand new cloak, and looked around for somewhere to hang it. His new office was small, and currently bare but for a desk, a chair, and some fitted shelves. On the desk was a pile of blank paper, an inkwell and some quills. After some thought, he folded the cloak and put it on the shelf.

Home from home. Literally, my own place on Tycho's staff. Now, how would I go about getting a few pegs put up? I'll ask Chris, he's just down the corridor. I wonder what this room used to be?

Right, what to do first? Old habits die hard – if in doubt, make a To-Do list.

He sat on the worm-eaten chair, which had probably been in storage until yesterday.

Oh, the top sheet of paper isn't blank after all.

He smiled to himself as he recognised Tycho's looped handwriting, pulling the sheet in front of him and reading.

My dearest Johannes,

I trust that you find everything to your satisfaction. But remember! Before you start your work on Mars, or on your Mystery – remember what you have promised me!

Yours in huge and comfortable wealth, a tiny pinch of humility, and considerable amusement, T.

He couldn't help laughing and shaking his head. He had been so angry when he wrote that letter, angrier than he had ever been, and the memory of it was still unsettling.

It worked, though, it worked! Maybe Tycho needs someone to stand up to him from time to time, and I'll do that if it gets us nearer to the Truth.

The young man took another sheet from the pile, and bent over it carefully, the roughly-cut quill scratching as it caught the paper.

Against Ursus. A defence of the works of the Imperial Mathematician Tycho Brahe.

This might take some time.

Notes on Chapter Sixteen

Kepler

Fictional characters:
None

Johannes Kepler was almost a polar opposite to Tycho Brahe. Their meeting was pretty much as described in the chapter (although he probably didn't faint). Their working relationship was no less fraught than their initial encounter; Kepler frequently stormed out over pay and conditions, but things were always somehow patched up. They disagreed over astronomy, mathematics, religion and philosophy – but both were passionately convinced that they were on the path to discovering the truth, and realised that they needed each other's skills.

Kepler's *Mysterium Cosmographicum*, which he sent as a calling-card to the great astronomical minds of his time, is an odd little book. He argues that the distances between the planetary orbits can be linked to the regular mathematical polygons, revealing an underlying harmony in the universe that is a sure sign of God's handiwork. This is one of Kepler's hallmarks – a thoroughgoing combination of theology and detailed mathematics that reminds us of the Ancient Greeks.

He was never afraid of the unconventional or of seeing things from an unusual viewpoint. In a technical and imaginative *tour de force*, he published a story called *The Dream* explaining exactly how the heavens would appear to an observer living on the Moon.

17

Humility

AD 1604

In which Kepler has succeeded to Tycho's old job, and attempts to keep his own ground-breaking research into planetary orbits going despite all of the demands that life in an Imperial Court places on his time and his emotional energy.

The Imperial bodyguards were always suspicious of John Kepler, but they were only echoing the general sentiment of the court. Although they understood why the Presence Chamber had to be kept locked and barred during the weekly astrological briefing, it didn't make them any happier to leave their Emperor alone with the saturnine young man. If he had tried, Kepler couldn't have struck them as more sinister, with his quiet voice, his limp, his penchant for archaic black clothing, his neatly pointed beard... and those eyes, that never quite seemed to fix on anything.

It would have amused them to know that Kepler himself found the whole experience equally awkward and distressing. Now he stood in his typical pose of humility, eyes downcast, hands twitching as he forced them by his side, tugging gently on his cloak. The silence had lasted even longer than the last time; had Rudolf fallen asleep while reading the report? It had happened before, and he hadn't known what to do then, either. Finally, he cleared his throat, but didn't dare to raise his head.

"Does my report please Your Majesty?"

He could still his tongue while awaiting the reply, but not his mind, that endless commentator. *Do other people have to put up with this incessant self-questioning?* he wondered, for about the thirtieth time that day.

And did my report, my pedantic, formal, report, please the Emperor? The million-thaler question. Although I think he has come to trust me enough that he would accept even unfavourable interpretations of the aspects. Or has he? He's a hard one to read, alright. And so here I stand, as I do every week, thinking pretty much the same thoughts as I do every week. My mind is revolving on its own little epicycle as I rotate around the sole source of light in my life. Or at least of gold, which helps keep the dark out just as well.

He suppressed the inevitable chuckle.

And every time I stand here I make that same little joke to myself, and just as I am about to put together a devastatingly memorable pun about eccentricity he interrupts m—

Rudolf pulled himself upright in his uncomfortable throne. He stood (or sat) on ceremony even when alone. His deep bass-baritone was confident, as always.

"Well, I shall look it over in more detail when I have time. Although your forecasts are much harder to read than Tycho's ever were."

Teeco. He pronounced it Teeco. If he were alone with you he would correct even the high-and-mighty Emperor, I'm sure. He certainly crushed my stumbling efforts at his outlandish pronunciation, but then, when we first met there was no reason for him to be well disposed. But what on Earth do you mean? My handwriting? But you've forbidden me to use a scribe for the weekly horoscopes, state security, and it's far too late for me to change my hand. It's a reflection of who I am. Cramped, cribbed, poorly formed, spidery, unconfident. How unlike Teeco's wonderful flowing script, you're thinking. Who is this nobody, you're thinking, appeared in court as another one of Teeco's clerks, and within two years the old man is dead and his followers are fighting for the job and this... mongrel with no family ends up Imperial Mathematician? But you appointed me, maybe just so you could criticise me more easily, I'm just a plaything of yours? Or maybe you mean my Latin. But I've sweated blood over my style, if I don't look or speak like a courtier at least I can compose like one.

Now you're waiting for me to say something. I'm meant to know what your thinking, I'm your mathematician for heaven's sake, your predictor, your augur. No, that's not right, the augurs dealt with birds and beasts. Where could I find out precisely when the Romans started to accept astrology? I think Vitruvius has something, and there's always Pliny as a last resort, but... no time for that now. You're waiting for me to respond.

"My handwriting, Your Majesty?"

The Emperor frowned with impatience. Such a hard man to read, even to another such.

"No, man, who cares about that? I only care for the essence. The hidden causes."

Oh. Oh no, he's going to ask about his sun-sign, and constellations, and the planets' positions on the zodiac, and I'm going to get angry and he'll cut my wages. Again. Or at least cancel my promised raise which is the same thing, and we're already going hungry since I spent all last month's stipend on this cloak so they wouldn't laugh at me. I've got to bite my tongue. I do that so often it's a wonder I have a tongue left. And with Teeco gone there's nobody left to argue this out with. Chris is still obsessed with draconitic precession, and Rothmann's too far away and His Majesty doesn't like his servants working with his rivals and anyway he's dying. And as for Galileo...

Calm down, John, calm down. One step at a time.

Rudolf looked hard at Kepler, who did his best to avoid eye contact as the courtiers had explained.

"Tycho would tell me when Mars was in Virgo, things that

made sense, things I learned from my tutors. Good old Sacrobosco; now, *there's* an author to admire. You shouldn't try to teach an old dog new tricks, young man."

The astronomer flinched.

"But, Your Majesty…!"

Did I just say that? Did I just object? But, after all, why not? Isn't this Prague, the centre of the civilised world? Isn't this Rudolf, the self-proclaimed patron of all the arts and sciences? Aren't I the Imperial Mathematician, paid to be brilliant and to think up new tricks? Teeco wouldn't have hesitated. Even Chris wouldn't. Why didn't the job go to him?

But then I'd be a starving nobody, and at least now I'm a starving somebody with a nice black cloak and a beard in the latest, ridiculous style, and the castle guards laugh behind my back instead of in my face. Deep breath. Think of what Teeco would say.

"But, Your Majesty, think what else you learned from your tutors. The world is changing, on earth and in the heavens. Sacrobosco's been dead these four hundred years —We're in the seventeenth century and who better to lead reform than Your Majesty, above and below... umm... 'As above, so below. As below, so above.'"

Damnation, that made no sense at all. And where did that Hermetic quotation come from? You know you're just pandering to his crank tastes, like Teeco did at the end. Why not tell him the real reason? That astrology doesn't work like that? After all, constellations are man-made marks in the sky: there are no lines in the heavens joining up the stars. And the stars are so far away they couldn't possibly influence us, because if they were closer you'd see them move as the Earth orbits, and...

Oh, I don't tell him this because it's dangerous politically, of course. Even Rudolf is bright enough to see the comparison with borders and countries. I think I've earned the right to keep my ideas out of politics, even if politics won't keep out of my ideas.

The heavyset man smiled, slowly.

"Bravo! That's the spirit! But maybe you could tell me what signs the planets are in, just to keep an old man happy?"

"Yes, Your Majesty. Of course."

Oh, well, if it "keeps him happy." It's not as if it's any more work: what does he think I do all day except precisely that? Boring boring boring, sometimes I think my head will fall off with the sheer tedium, and time is so precious. When I'm in precisely the right place and the right time to understand the way the universe works! When I name my navigation tables after him he'll see it as a compliment, not a perpetual reminder of what he stopped me doing.

Maybe I complain too much. Things aren't really that bad. It must be my destiny to complain, I was born under a particularly unfortunate aspect of planets.

"Was there anything else, Your Majesty?"

Rudolf considered the youngster in front of him — too young for an Imperial Mathematician?

"You're always very keen to leave me, aren't you? Just like Tycho! Why are my other philosophers so different, hmm?"

Well, in my case it's because I'm frankly terrified by the whole experience. I dream sometimes that all this is a nightmare. Hang on, that can't be right. Maybe I mean that I dream I'm awake? No...

Sometimes I fear that I will awake and find myself back in a hovel, with no father and coughing my lungs out. Everything has happened too quickly. And all because geometry is so much simpler than humanity. But Teeco? He wanted to escape these confidential interviews? That's news to me.

Of course, he distrusted astrology because he couldn't measure the planetary influences, but that never stopped his flow of predictions. Or the trouble they got him in! Ahh, Teeco, Teeco: now I have it, you old dog! You were uncomfortable with the one person in the country with more power than you. The magic touch might not work on him, mmm?

Maybe you did learn your lesson from Denmark after all, even though your disguised your exile as a Royal Progress.

I never knew what you saw in me. Richest man in Christendom and his poor apprentice: you never believed in my Platonic separations and I openly despised your so-called planetary system. But you named me your successor, and that weighed more with His Majesty than Chris' claim as your right-hand man, or the family protests.

It pleases me immensely that I've found a weakness in the armour, one little instance of the Almighty Teeco not getting his way. Maybe we're more alike than I thought. Why does that make me even more terrified?

"I suppose... that mathematicians prefer numbers to people, Your Majesty."

Or at least to ciphers.

Rudolf's smile returned, and he rested the scroll on a table to his side.

"Hah! That reminds me of something clever I heard..."

Please, not Galilei. My tongue hurts too much already.

"This Italian fellow, says that Mathematics is the language of the Book of Nature. Oh – and that the Bible tells us How to Go to Heaven, not How the Heavens Go! Isn't that good?"

Kepler could contain himself enough that his muttered reply was icy rather than angry.

"I believe that Plato was Greek, Your Majesty, though Cardinal Baronius is certainly Italian."

He half hoped that Rudolf hadn't heard.

"No, they weren't the names. Now, what was it? I have it on the tip of my tongue – ah, yes, Galileo."

"I believe that Signor Galilei has been known to borrow quotations without due credit to their authors, Your Majesty."

Although at least he didn't steal the theories, not like Ursus. Now, there's a story and a half: pity he died before I could get my book about it out. Not that anyone else could have come up with Galilei's half-baked theories. His observations are not only unsupported by any coherent optical theory, but he's proud. Hubris calls for nemesis to strike him down! And not proud like Teeco: he was proud because he could always defend his ideas and he was consistent! Whereas this... Italian... upstart... changes his argument five times a day and is still wrong.

I even tried to help him. I wrote him friendly letters and he didn't reply – instead he publicly attacked my theory that the moon causes tides! When he can't even get the right number of tides per day!

I wouldn't mind if he were just another nobody. But he's well connected and when he finally pulls the Church's nose too hard he's going to get into trouble. For over fifty years astronomy has been the only area where Catholics and Protestants have worked literally side by side and I don't want that to stop.

Besides, it's demeaning to steal quotations.

"Ah well, that's the world of scholars for you. But where would we monarchs be without you, mmm?"

Excellent question. Why do you need me? Us? Your ever-growing retinue of academics who couldn't hold down university appointments? Myself included, I suppose. Vanity? One-upmanship? I would be willing to accept genuine intellectual concerns if I started to see more evidence. Should I say something? No, he's only pausing for dramatic effect, doubtless he thinks he's got some crushingly witty comment.

"More to the point, where would you scholars be without us?"

In my study doing proper work. I always feel I'm running out of time and all this small-talk is just so slow. Court functions are worse, but if I scowl enough and sit in corners people tend to leave me alone.

How could he stop himself from sounding sarcastic, in the circumstances? There was a grain of truth here, after all.

"I am eternally grateful for your generosity, Your Majesty."

"Before you go... I assume you can keep a secret?"

This is bad. This is very bad. I'm not the sort of person who gets trusted with secrets. Which is a pity, since anyone would agree that I'm naturally secretive. Which, of course, makes me "sly" and "untrustworthy." Circles, circles everywhere.

Anything which the Holy Roman Emperor wants kept secret is definitely something I don't want to know about. But on the other hand, the forecasts I write for him are already secret. Is trust automatically symmetric? There's no way I can possibly tell him what he should or shouldn't tell me, though. Honesty is the only policy that ever seems open to me.

"I am not known for careless talk, Your Majesty, and nobody yet has had cause to doubt my word."

Is that strictly true? I wonder. I would so hate to say anything that isn't strictly true.

The Emperor leaned forward, and Kepler smiled at the cheap theatrical effect. "Good. Then... suppose... I were to take steps against my brother – don't look like that – you would be able to find a suitable interpretation of the stars..."

Planetary aspects, planetary aspects.

"...that could be published, showing my actions to be justified in the eyes of God?"

Plato is dear to me, but dearer still is truth. I bet Galilei would even claim that one for his own. Alas for the shade of Aristotle. And alas for the Noble Art. I'm beginning to realise quite why Teeco could be so cynical.

"Your Majesty! I am well aware that some astronomers, even — forgive me — previous Imperial Mathematicians, have treated horoscopic predictions as a joke or an elaborate fiction. If for one *instant* I believed that you were of that opinion, or encouraged it in your court, I would have no option but to resign my post."

Oh, wonderful, what a threat. And be sent back home, into the middle of a war-zone to have my life destroyed yet again by one side or the other or both for refusing to sign up with them. Sometimes I really passionately wish that I didn't care as much as I do. Other people manage to bend their principles, I see it all the time, this whole court is based on that phenomenon, so is the Empire. Maybe I've gone too far this time.

"Your work means that much to you?"

At last, a question with an easy answer!

"It means more to me than anything else, Your Majesty."

"But how can mathematics mean more to you than your Emperor?"

Because — because mathematics is how God expresses himself in the Universe. Because when we understand the behaviour of a function, we know everything about it: we think the same thoughts that the Almighty does and that is why the Bible tells us that we're made in His image. We are only rational insofar as we are mathematical — rational, ratios, everything is in an orderly ratio to everything else, not just numbers and shapes and sounds but the spheres of the world, all in harmony. Because God reveals Himself through geometry, and the Holy Trinity can be seen in the elegant definition of a sphere. Because the physical world is built of numbers, and not in a mystical sense like Pythagoras thought, only to be understood by initiates, but through the spiritual harmonies which were before Time and persist for ever, lying within the grasp of anyone who wants to stop and think. Because mathematics shows us beauty and power and fate and freedom and elegance and clarity. Because — because numbers make sense, and nothing else in this increasingly crazy world does. Because Rheticus the Lutheran worked alongside Copernicus the Catholic using techniques from al-Tuci the Mohammedan, and because Teeco the nobleman worked alongside me, a pauper, and however briefly we made the world see itself in a new light. Because I follow where Teeco left off and he picked up from Copernicus who followed Regiomontanus who relied on Peurbach and all the way back to Ptolemy, no, further, to Callipus and Aristotle, to Eudoxus, to un-named Babylonians, and, and, because it's only through mathematics that the we are slowly pushing back the veil of darkness and ignorance surrounding us, and because now the flame is flickering and Emperors at a whim can make me write horoscopes when I'm this close to finally understanding.

Ah well, when we've spiralled down into the pit at least we can trace our trajectory and smile at Archimedes' foresight.

"I... I don't know, Your Majesty. Merely a figure of speech."

Or a speech about figures?

Rudolf sighed, leaning back, seeming older than his fifty-odd years.

"Very good, Kepler, you're dismissed. Ask the guard to summon the Imperial Alchemist on the way out. Same time next week, and remember to tell me where Mars is, hmm? And think about what I have said."

The young man bowed his way out, trying to work out whether this meeting had been a success or a failure.

"Yes, Your Majesty. Thank you, Your Majesty."

Whom do I detest more for putting me in this situation? Myself, Rudolf, Teeco, fate, God?

I complain too much. An inauspicious aspect of planets at my birth. It would appear it's time for the next repetition of the cycle...

He turned to leave, hoping that the hammering on the door to summon the guards covered his sigh and that his face was low enough that they would not see his sudden blush. On the whole, more of a success than a failure. But how long could he sustain the balancing act? Back to his study, his retreat, his own fortress against the changing aspects of his own system of influences.

Same time next week.

Notes on Chapter Seventeen
Kepler

Fictional characters:
None

Tycho died due to complications arising from a burst bladder – he drank too much at a banquet and was enjoying the conversation too much to leave to relieve himself. (At least, that's the accepted story. Some historians blame mercury poisoning, either from his alchemical researches or as murder, sometimes blaming Kepler. This seems a step too far for the cautious and paranoid young astronomer. Recent research, including exhuming Tycho's corpse, supports the burst-bladder theory.)

Kepler inherited Tycho's position, in the absence on sabbatical of Longomontanus (Tycho's longest-standing assistant). He had to fight to get hold of the observational data, which was claimed by the Brahe children. Once he had access to all the measurements, he began his 'war with Mars' to establish precisely what its orbit was. After years of work, and many false starts, he established that the planetary orbits *cannot* be broken down into combinations of circles. Instead, a different mathematical shape, the ellipse, must be used, overturning the consensus that had held since before Aristotle's time. Several historians suggest that this 'Keplerian revolution' was more important than the Copernican.

He was a keen amateur historian himself, and pieced together Osiander's contribution to Copernicus' work (the infamous *Preface* had been anonymous and was variously attributed).

Kepler's thoughts in this chapter are largely based on the

personality that shines through his surviving diaries, which are frank and often self-deprecatory.

18

Pride

AD 1600

In which we step back in time just a few years, and cross the Alps to Rome, where a troublesome priest called Giordano Bruno has been stirring up huge controversy with his radical theories about astronomy and theology, seen from the viewpoint of Robert Bellarmine, a leading light of the Church.

January 16th

Your Holiness,

As requested, here are my thoughts *in re* Father Giordano Bruno, strictly off the record. You have, no doubt, the full file before you. If it were ever to be bound, it would run to several handsome volumes already. When a man has been a prisoner, and questioned regularly, for seven years the paperwork is necessarily immense – you know how we operate! I can understand why you wish me to summarise matters, and am very conscious of the honour you have bestowed upon me; some of the other judges in the case are officially senior to me.

I am also fully aware that this is a breach of protocol on both of our parts, and am writing this letter to you with my own hand in the expectation that it will be destroyed once read. Strictly speaking, as the prisoner has appealed directly to you, I feel ambivalent about giving you any *advice*. I must stress that everything that follows is my own personal private opinion as Roberto Bellarmino, your old friend, theological consultant, and confidante, and not in any way representative of the views of the Court. As Cardinal Inquisitor Bellarmino, I have pronounced the man guilty (pending appeal to the Pope). As Roberto, I am not quite so sure.

You will be aware of the awkward political situation surrounding the Bruno case. We have to consider the international reputation of the secular states of Rome and of Venice as well as that of the Papal Curia. The man was handed over to us by the Venetian Inquisition, in direct defiance of the political settlement reached over the course of many years and much hard diplomacy with the Serene Republic. It has proved in all of our best interests to maintain a state within the Italian peninsula where the Ecclesiastical Authorities have been considered to have a lighter touch than elsewhere. Heresy can be contained and observed, and the famously liberal Venetian printing presses attract many, from both sides of the Alps, who wish to propagate novel ideas of all kinds. Some of these may prove useful to Mother Church in due course, and it is greatly to be desired that they be encouraged in a territory that is within an arm's easy stretch of Rome.

Similarly, we must consider what is possibly our trump card very carefully. If you do not allow Bruno's appeal, and confirm our Guilty verdict, then we are done with the matter. We will have formally found him a heretic; in this particular case, it is the civic authorities of the City of Rome rather than the Curia of the Church that will pass sentence. Although we may suspect that the consequence will be burning alive, with or without various tortures beforehand, we cannot actually *know* that, since the civic authorities operate their own due process of justice. We are simply ruling on a theological question referred to us by the Venetians. It is purely an accident that it happens to be a crime in the City of

Rome to be a heretic!

I can see you smiling and sighing. You're right, this is so much hand-washing. If you disallow the appeal, the man will die, and he will be on our consciences. Is he guilty of heresy? Certainly. Should he die? On that point I cannot decide. In making your decisions, you will need to know what sort of a man he is, what sorts of things he is saying, and what will be the consequences of your decision. Please allow me to tackle each in turn. (You know that I am a systematic – some would say plodding – thinker, and such a comprehensive strategy appeals to me. I beg your indulgence.)

Bruno is an unassuming little man from Naples, of about fifty years. He would pass unnoticed in a crowd, being of small stature with no distinguishing marks save for a well-kept beard. He joined the Order of Preachers as a youth and was ordained priest; until his arrest, he wore clerical or secular garb, quite contrary to all good order, as he saw fit. He learned well, being a ready scholar of Thomism. He became famous as a young man for his incredible feats of memory; you have quite possibly read some of his (perfectly innocuous, and indeed useful) booklets on the subject. There were questions raised early in his career – he has freely confessed to reading prohibited books throughout his life, but given his gifts, he was predicted to go far.

He did go far – all the way to the Protestant states of Switzerland, to France, to various of the German-speaking lands, and even to England. There he mixed with like-minded academics and heretics (often the same in those countries) and repeated a pattern several times. He would demonstrate his (undoubted) rhetorical and intellectual prowess, haunt the cloisters of a prominent university, acquire a rich and powerful patron, and then launch into intemperate attacks on all conventional wisdom and on the social order of whichever country was his host at the time.

His patron would drop him like a hot iron, Bruno would feign indignation and rage, and leave for another place only to start the same sequence once again. Each of his patrons flattered himself in turn that he would be more tolerant and protective than the last, which inspired our Bruno to greater heights of passion in his ill-natured controversies. To this day, after seven years as our very special guest, he maintains an attitude of bluster and self-importance which would shake the confidence of a weak man, no matter how devout.

He has worn various guises throughout his travels; the Dominican friar, the scholar, the man-of-the-world, even the Protestant theologian. He is very proud of having been excommunicated from at least one sect which he had never even joined! In short, he cannot hold a conversation for more than a few words without proposing something offensive and ridiculous; he gathers enemies as naturally as a dog gathers fleas.

This is highly unfortunate. In these cases, it is hard to distinguish the man from the ideas – the sinner from the sin. We must be absolutely sure that we are punishing the ideas, the sin, rather than the man himself, which would be so much easier with a calm and even-tempered person on trial. A bellicose and destructive nature is not in itself a crime or even a sin, if properly directed. Indeed, naming no names, many of my colleagues in the Holy Office might seem to an outsider to be remarkably similar in personality to Father Bruno... whereas poor, systematic, plodding Bellarmino could never be accused of rage, merely sadness.

So what are Bruno's ideas? Once again, we must be very careful to distinguish between the various strands of his thought. It is entirely possible that a guilty verdict will send out the erroneous message that *everything* Bruno claims which is not explicitly approved by the Church is therefore heretical. I do not think that we can claim this. As your friend Roberto, as your theological adviser, as the author of the *Disputationes de controversiis*, and as Cardinal Inquisitor I can safely affirm that there are large areas of human knowledge which are completely neutral, neither required nor prohibited as beliefs. We must pick our battles carefully, and avoid being sucked into petty and insoluble arguments about the content of the no-man's-land between the two, being content to map its borders clearly.

In the essentials, he maintains that as God is Infinite, it would demean Him to have a Creation which is only finite. The universe, therefore, must be infinite both in space and in time. If the universe is infinite, it follows that there must be an infinite number of beings therein. Since they are not here on Earth, there must be other Earths containing other beings. He affirms that some of these Earths will not have experienced Original Sin, as their Creations resisted Satan's temptations. But some of them, such as ours, fell: *and on each such world there must have been a crucifixion and redemption.*

He develops other theories to explain how his Universe of Many Earths will behave. Casually rejecting Aristotle and all his successors, he rules that there can be no centre to the World (since it is infinite) and therefore there is nothing to distinguish the earthly elements (earth, water, air, fire – attracted to the centre of the universe) and the heavenly element (aether). He therefore claims that all five elements must behave in similar ways. The Heaven being nothing special, it is no longer God's home. God is omnipresent throughout His Creation and coextensive with it (since both are infinite). Christ Himself is not unique (being replicated on every fallen Earth) and thus is only a mortal being, chosen by God – the old Arian heresy revived.

Where are all these Earths? Why, on spheres around the stars (except he also rejects the spheres), which are nothing other than different suns, very distant from us. It follows that Earths (including our Earth) must circle their sun-stars, rather than *vice versa,* and he adopts the system of calculation developed by Copernicus some sixty years ago.

Some of this is harmless enough. Cardinal Nicholas of Cusa maintained the plurality of inhabited worlds without thereby falling into the theological error of reasoning that we can deduce facts about the nature of God and of Christ from the mere possibility of there existing intelligent races other than humans and angels. The learned Cardinal also speculated about infinity, as have other perfectly respectable theologians, but did not go so far as to suggest that this would force us to re-evaluate the nature of Christ.

Similarly, we do not have an official position on the Copernican controversy, and have remained silent upon it for sixty years, for the very good reason that nothing theological or spiritual rests upon it. It is entirely possible, indeed likely, that our Hebrew forefathers believed that the Earth was flat and relatively small, and we find echoes of this belief in the texts which we have. If our philosophers were somehow to prove that the Earth moved and the Sun was stationary, it would not be difficult to interpret the Bible accordingly. Personally, I do not think we shall ever see such a proof, but doubtless the Ancients said much the same of finding the size of the Earth before Eratosthenes came along. Copernicus was a faithful member of the Church, and his idea's popularity amongst the Protestant universities should neither encourage nor dissuade us. (I am led to believe that even Luther would have agreed that two and two are four.)

I wonder, I genuinely wonder, how much of all this Bruno actually believes in his heart. In conversation, even during his trial, he is slippery. He never defends a position squarely, he attempts to evade questioning by challenging the definition of words, by asserting that *properly understood* he agrees with the doctrines and teachings of the Church – but it is he, and he alone, who is entitled to define exactly what the proper significance of words should be. It reminds me of my Jesuit training, when the so-called Socratic Method of questioning was unleashed upon us! (Which, in turn, reminds me of another fact about our accused. Bruno *also* claims to have translated various works of Plato, but has merely read Ficino's elegant, if inaccurate, translations and altered some of the words. Although I am ashamed to say it, as a classicist myself this seems almost as bad to me as his theological and philosophical transgressions.)

So much for the harmless nonsense. Wrapped up with these speculations, wrapped so closely in his mind that he cannot see that they are really distinct, are his grave theological errors regarding the nature and person of Christ. If he were a layman, I would be inclined to turn a blind eye. If he were a quiet scholar, I would be inclined to ban his books but leave the man alone. As it is, he is an ordained priest and a self-admitted controversialist, and he is in our hands. He cannot plead ignorance of the seriousness of the charges against him.

He is probably right to feel frustrated by the surprisingly hard line taken on him by the Venetian Inquisition. He had appeared in Venice to repeat his regular pattern of academic and social controversy, attempting to insinuate himself into the universities and to find a patron, while overseeing the publication of certain of his pamphlets. I rather wish that this troublesome friar had remained safely across the Alps, or even that my Venetian colleagues had sent him packing northwards rather than referring the case on to us. But he's here, and to my mind there is no possible verdict other than Guilty As Charged. His appeal to you is within his rights, but is obviously a delaying tactic.

I said, when I started writing this letter to you some hours ago, that I should not give advice. Instead, let me once again be systematic. It's in my blood.

Possibility 1. You allow the appeal. Bruno goes free, flees Italy, and continues his career. All those within the priesthood of the Church who have been harbouring heretical views but keeping them quiet come crawling out of the woodwork, waving pamphlets, and we find ourselves being mobbed by a thousand pygmies.

Possibility 2. You deny the appeal but allow Bruno a chance to recant and return to the bosom of the Church – and he accepts. This is the best possible situation, and is in fact likely (although less so than the possibilities below). He is aware that a failure to recant will almost certainly prove fatal to him. I believe that a man so self-important will also find his preservation important.

Possibility 3. As above, but he refuses to recant. The City of Rome sentences him to something other than death. This is almost as bad as Possibility 1. Bruno will be seen to have faced down the Roman authorities, and while I personally care less for the Mayor of Rome than for its Vicar, relations between the two should be kept as harmonious as possible.

Possibility 4. As above, but Bruno is burned at the stake. Given the situation that we are in, this is probably the most likely outcome. The Holy Office and the City of Rome retain their reputations for being effective arbiters of justice, unafraid to do what the laws of God and of man require. *However*, we start to put ourselves on the moral level of the Protestants, whose first reaction to anything seems to be to burn something or someone. *Also*, unless we are very clear about the exact specifics of Bruno's heresy, we risk observers considering that some entirely unexceptionable opinions are in fact inextricably tied up with heresy, and thereby discourage loyal and devout Churchmen from pursuing profitable avenues of enquiry.

It is not giving you advice to tell you, as a matter of fact, what I would do if I had to choose between the various options currently available. It is, simply, a statement of how I see the situation, and I am aware that you will also consider many other factors and, indeed, will have deeper insight into matters of which I am only vaguely aware. Having said all that, I would find Bruno guilty but give him the chance to recant and repent. If he did not accept this, I would make a great show of urging the Roman authorities to be lenient and to respect a philosopher's right to speculate, while secretly ensuring that the man would be burned as soon as might be arranged (and in this perishingly cold winter, any fire would draw an admiring crowd).

With all my continued prayers and devotion, Roberto Bellarmino

February 18th

Your Holiness,

I was present, *incognito,* at the burning of Giordano Bruno yesterday. You have doubtless received the official reports, or if not they are surely on their way to you through the correct hierarchical and bureaucratic process.

Although you know that I was – and am – fully in agreement with the sentence of the City Court of Rome, I was troubled by my mis-reading of Bruno's character. I had spoken to the man, on and off, for months. I had read transcripts of his conversations stretching back years. Yet I had no idea that he would hold to his ideas so strongly. He was insisting on preaching to the crowd even as they piled the wood around him; eventually the constables were obliged to clamp his tongue with wood to prevent him from speaking. This displeased the crowd, and I wondered whether to intervene but let it go.

I count myself experienced in reading the personalities of men. I suspect that I made the big assumption that Bruno was rational, that he would behave as an academic or a merchant or anyone else who cared for the future and thought about the present would behave. Was he, after all, simply mad? It takes many different forms.

It also made me think about the more general fight against heresy. Hearing the chatter of the crowd, all that they cared for was the fact that he was "a dirty heretic". None of them seemed to stop to consider whether their own beliefs were entirely orthodox. I am convinced that under sufficiently strong questioning, most of them would turn out to be at least partly confused about the elements of their faith, and yet, to them, heretics simply seem to be a breed of monsters, bogey-men who doubtless eat babies and rape young girls (or *vice versa*). So much for my worries about the subtle distinctions of ideas.

It is far too early to say what the reaction will be amongst our real audience, who will be hearing the news over the next few weeks, months, years; the priests and scholars of our own lands and, yes, those who have split off diplomatic relationships and broken the Communion of the Church too. Surely *they* will realise that this is as much a matter of internal Church discipline as one of theology, and certainly not a judgment upon strictly philosophical or astronomical ideas.

We shall see.

R. B.

Notes on Chapter Eighteen

Bruno

Fictional characters:
None

Giordano Bruno was an odd combination of genius and charlatan, scientist and mystic, visionary and madman, charmer and megalomaniac. It is hardly surprising that his presentation in history books depends mainly upon the inclinations and sympathies of the historian concerned. His story touches on many of the themes raised by the Protestant Reformation and the Catholic Reformation (today's preferred name for the Roman Catholic Church's attempt at putting its house in order – the old name of the Counter-Reformation is seen as misleading, since some strands of this movement started well before Luther appeared).

Throughout the seventeenth century, certain elements of the Roman Catholic authorities attempted a subtle and nuanced approach to questions of heresy and of Papal authority, as exemplified by the profoundly intelligent and thoughtful Robert Bellarmine (later canonised). For many reasons, some touched on here, this policy either backfired or was misunderstood by everyone outside the central decision-making bodies of the Church. Perhaps academic subtlety was not received well when Europe was tearing itself apart in a series of religious wars, rebellions and conquests.

Bruno's execution sent shock waves through academic circles. Prominent authors in Catholic countries simply stopped writing.

Notes on Chapter Eighteen

Bruno

Principal characters

None

Giordano Bruno was an odd combination: a genius and charlatan, scientist and mystic, visionary and madman, charmer and curmudgeon. It is hardly surprising that his reputation in history books depends mainly upon the anachronistic sympathies of the historian concerned. His story touches on many of the themes raised in the Protestant Reformation and the Catholic Reformation (this is the term now used by the Roman Catholic Church's attempt at putting its house in order: the old name of the Counter-Reformation is seen as misleading, since some strands of this movement started well before Luther appeared).

Throughout the seventeenth century certain elements of the Roman Catholic authorities attempted a subtle and informed approach to questions of heresy and of Papal authority, as exemplified by the profoundly intelligent and thoughtful Robert Bellarmine (later canonised). For many reasons, some touched on here, this policy either backfired or was misunderstood by everyone outside the central decision-making bodies of the Church. Perhaps academic subtlety was not required at a time when Europe was tearing itself apart in a series of religious wars, rebellions and conquests.

Bruno's execution sent shock waves throughout academic circles. Prominent authors in Catholic countries simply stopped writing

19

Arrogance

AD 1616

In which we meet another of the famous names in our story, Galileo Galilei, and, although we learn that his staunch defences of Copernicus' views were (mostly) correct, we also find him to be a prickly character who did himself no favours with his stubborn personality.

AD 1616

My dear Signor Galilei,

Many thanks for your letter of 15th April. I would be entirely happy to do as you have requested, and to issue a public clarification of the dealings between your good self and the Holy Office. I hope that you will not object if I take this opportunity to expand on my own remarks at our last meeting; it is often easier to express oneself in writing.

Firstly, I am aghast that rumours of a conversation about suspected heresy are being spread about you. I believe that you are accumulating some enemies, such as those who referred you and your works to the Office for clarification and guidance! It will be a great pleasure to explain to these paranoid professors that not everybody who passes through my office door is there to be reprimanded. I enjoyed our frank discussion about the relationship between scientific enquiry and the doctrines of the Church, and believe that I have helped to clarify matters for you. It is good to put these things on paper, so I repeat that we have no objections whatsoever to the use of the heliocentric (or "Copernican") model of the world, so long as it is stressed that it is precisely that, in other words a model, a hypothesis, a theory.

As I have previously commented (and I believe that the comment has been widely circulated), if proof of the Earth's movement were to be found, then the Church would of course accept it. However, to my knowledge, no such proof exists (and we may doubt whether it ever will). The Bruno affair rather forced our hands and required public clarification of the Church's position on the controversy. Did you ever meet Bruno? He applied for the chair of mathematics at Padua at about the same time as you, and you must at least have had friends in common. He was an impetuous, impulsive, infuriating little man – if he had been as reasonable as you, the case would have ended very differently.

In the absence of such proof, the Holy Father has decreed that it will be an error of faith to defend or maintain the Copernican hypothesis as the undoubted truth. I have in front of me your signature on the document by which you swore to abide by this teaching – I affirm once more that there has been no suggestion, in any way, by me or any of my colleagues, that you have so defended or maintained the hypothesis in that manner up until now, contrary to the accusations of your enemies.

It is but rarely that we find it necessary to make rulings upon what may or may not be maintained concerning the physical world. Indeed, certain of such pronouncements in the past were proven to be spectacularly misguided and were subsequently overturned – you will guess that I am referring to the so-called Condemnation of Paris. You will also understand that we are not treading on the toes of new enquiry, rather attempting to prevent philosophers from falling into error by asserting as truth that which has not been proven. Truth comes from God, and all arguments eventually reach it.

What is really at stake here is not a point of natural philosophy, and I would be grateful to you if you could communicate this to your esteemed colleagues at the *Accademia dei Lincei.* My concern is, of course, to establish and assert the right of the Pope to rule on any matter and for his authority to be recognised in matters temporal as well as spiritual. (You may, if you wish, read this letter to your fellow Academicians – I shall have no objection to any part hereof being made public.) The sad burning of Giordano Bruno was one of many factors which combined to spark off the late Venetian revolt against Papal authority and the subsequent re-statement of our position. We devout Catholics must recognise that the Church has not only a right but a duty to intervene in all aspects of life (including the intellectual when it leads us into heresy). Christianity does not stop at the church doors!

If you wish to discuss any of this further, I would always be available to see you. I have followed your distinguished career with interest and am always delighted by your new technological innovations.

With cordial good wishes, Cardinal-Archbishop Roberto Bellarmino

AD 1633

"So, have you heard about Galileo?"

"I heard they found him guilty – have they sentenced him yet?"

"Yeah, life imprisonment. Pretty harsh on an old guy like him."

"Harsh? I was expecting them to burn him!"

"Give over, they don't do that any more."

"More's the pity. The bigger they are, the harder they fall. He was quite a bigwig, wasn't he? All that telescope stuff up in Venice."

"They say he got paid a fortune for that. Military technology, and everything."

"Yeah, and him a professor. Who would have thought he could have been a heretic all this time?"

"I saw him once, you know, on his way to the Vatican. Nice clothes, upper class airs and graces – looked the part."

"You know he had three bastard kids before he got married? Couldn't marry bastard girls off so he packed them off to a nunnery. Shows you what he was really like."

"Yeah, I suppose. You never can tell. All that money, all that fame, and where did it get him?"

"Locked up in a nice villa somewhere, everything laid on for him. Not such a bad retirement!"

"Give over, you wouldn't like to have guards following you everywhere, would you? He'll never get published again, that's for sure."

"Did you read it, then?"

"What?"

"His *Dialogue*. The book that got him into trouble."

"Nah, you know I've got no time for books! If it's a good story, they'll make a play out of it."

"You fool, it wasn't a story. It was all about how the Earth moves around the Sun."

"I call that quite a story! So, you've read it, Mister Clever?"

"Not exactly. My brother's boss, he's quite a reader, and he was telling me about it when we had him round for dinner last Michaelmas."

"Who, the banker? What's he doing reading philosophy?"

"No, no, apparently it's not like that. It's a conversation between these guys, one of them saying the Earth moves, one of them saying the Sun moves, and one of them to keep score."

"Don't see as how a book like that could get you locked up."

"Yeah, well, me neither. But apparently old G was warned off this, ages back. Remember old Bellarmino, used to throw his weight around?"

"Sure! He was a real gent. I heard him preach once, when I was visiting my uncle up in Capua. Clever guy, and such a good speaker. My uncle told me that he wrote a letter to the Heretic King of England, correcting his Latin grammar! That's stylish. Surely he can't be still around?"

"Nah, died ages back. But he'd told G not to say he'd proved his damnfool theory, just to say it was... a possibility."

"Makes sense."

"Except that G reckons that he *has* proved his theory. It's all to do with the tides, apparently. You wouldn't get tides unless the Earth was spinning."

"Why not?"

"Beats me. Anyway, the judges said he'd got it all wrong. Apparently, what G says in the book would mean that there's a tide every day."

"And isn't there? I like to keep safely away from the sea, it's not natural to travel on boats!"

"No, you fool, there are two. According to my brother's boss, there's this famous German guy who says the tides are caused by the moon, but G says that sounds too much like astrology to him."

"What's wrong with astrology?"

"Well, precisely. Anyway, the judges tell our Galileo that he can't say it's proven when his theory is all up the spout. But that's not the half of it."

"Go on, what else?"

"Well, like I said, there's one of the people in the book who says that the Earth is in the middle of the world, just like we all know and like the Bible says, right?"

"With you so far..."

"So he's kind of the spokesman for the Pope, yes?"

"I suppose so."

"Guess what this guy in the book is called?"

"What a stupid question! How could I possibly guess that? Hang on, is it Urban, like the Pope? No? What was his name before he was Poped... umm... Maffeo? No? Go on, then."

"It was 'Simpleton'!"

"Never! Hell's teeth, that wouldn't go down well. You're having me on! I thought he was mates with the current Pope, even if he had run-ins with some of the old ones."

"Well, G swore blind that *Simplicio* was an old-time astronomer who said those things, back in the Roman Empire. He said he had no idea that the name also means half-wit. As if anyone would buy that!"

"Not surprised they locked him up. Now I *am* surprised they didn't kill him!"

"Yeah, well, they couldn't really, could they? He recanted."

"He what?"

"Said he was sorry, he'd got everything wrong, he didn't believe a word of it any more, and that he'd be a good little boy in future. In his nice safe prison."

"He never!"

"He did. So they're burning his book instead of him."

"Well I never! So, who're you backing in the race tomorrow?"

AD 1642

To the Most Noble and Honoured Grand Duke of Tuscany, His Grace Ferdinando II,

Greetings,

I acknowledge your letter of March 3rd, regarding the Galilei case. Although I can understand your wish to commemorate such a singular and ingenious member of your court, I am afraid that there can be no change to the decree issued by the Papal Legate to Florence. Quite simply, we will not permit the burial of the late Signor Galilei within the main body of the Basilica of Santa Croce, nor can we allow the planned mausoleum in his honour to be built.

You have been a good friend to Rome, Ferdinando, and deserve more of an explanation than the bald declaration from the Legate. Neither I, nor my uncle, take any personal pleasure from this move. Galileo Galilei was a good and consistent friend to our family long before any of us were elevated to the positions which we occupy today. I myself have fond memories of him.

We were delighted to agree to your request that he return to Florence for medical treatment after his blindness, and were more than happy to allow him to live out the rest of his sentence of house arrest amongst surroundings he would find congenial. I understand that for the last four years, he lived in some style and am grateful to you for any assistance, financial or otherwise, which you provided to him in his infirmity – a truly charitable act towards a unique man.

We do not wish to detract from his many achievements. Leaving his astronomical hypotheses aside for one moment, who else of our generation has contributed so much to optics, to mechanics, to timekeeping, to thermometry? His contributions to any one of these difficult and abstruse fields would be enough to justify him a monument raised by any patron with an ounce of appreciation, and this man excelled in all of them.

In astronomy, his discoveries are almost too numerous to mention. He discovered that it is profitable to turn a telescope skywards, and thereby observed the Heavens in more detail than anyone before him. He has convincingly argued that other planets have satellites which orbit them, he has described the mountains of the Moon, and he has observed the phases of Venus.

This, however, is where he fell into error. He claimed that these phases indicate that Venus must orbit the Sun, just as the phases of the Moon reveal its own particular motions. So be it. *It does not follow* that the Earth also orbits the Sun – my advisers inform me that many eminent astronomers have previously conjectured that the Sun orbits the Earth (in accordance with received and revealed wisdom), while the other planets orbit that great luminary. This avoids other complications; if the Earth moved, the fixed stars should, I understand, seem to change their positions contrary to all observation. I have no doubt that some such system will one day be proved by sufficiently precise physical investigation of some kind. His ridiculous arguments regarding the tides have been sufficiently discredited not to warrant further discussion here.

It is a pity that such a great mind overthrew itself. Had Galilei remained content to describe the new facts about the Creation which he had observed, all would have been well and there might have been a monument to him inside the Vatican, let alone Santa Croce. In passing beyond what he had observed to make bold claims which turned out to be baseless, he not only fell into an elementary philosophical error, but also into a theological error. He was very well aware that the heliocentric system can only be advanced as a hypothetical method of calculating planetary positions. Indeed, I have here in front of me a declaration to that effect signed by Galilei before the much-lamented Cardinal Bellarmino, who did so much to explain and clarify the limits of faith.

For all his brilliance, Galilei was not the only competent astronomer in Rome. Had he discussed his books with us before rushing to publication, we would have advised him of the ways in which his ideas outstripped the available evidence, and reminded him of the Papal decrees on the matter. Incidentally, we might also have advised him that certain elements of his book would cause great offence to one who had thitherto been his patron and protector, namely my revered uncle.

I have said that I admired Galilei, and that is true. However, his personality was – as you will be well aware – abrasive and combative. A sense of righteous indignation is all very well in its place, but humility and caution befit the scholar and the cleric alike.

Bearing all of this in mind, we cannot allow his burial in such a magnificent location, nor his commemoration in the manner which you have proposed. We have no objection at all to his being buried within the Basilica's sanctified precincts – he was, for all his faults and for all the irregularities of his personal life, a loyal son of the Church at heart. A mausoleum at state expense would do more than commemorate him, though. It would send out two messages, neither of which is acceptable to us.

The first is that heresy, in any form, is acceptable if balanced with genius. You may say that Galilei recanted of his heretical views, and you would be correct. Technically, though, he was condemned of violent suspicion of heresy, and although his confession and recantation may have saved his soul (and let us pray that is so!), it cannot erase that fact from the legal record. The more general point is that heresy becomes dangerous precisely to the degree it is mixed with genius. Nobody listens to the raving fool at the marketplace; everybody listens to the august and wealthy Professor, friend to the Holy Father.

Secondly, what we are really concerned with here is Galilei's direct challenge to Papal authority, as clearly laid down by my predecessors. This was obviously premeditated – one does not write a book on the spur of the moment! Such a move would send a message to Italy and beyond that Tuscany is at odds with the Papacy, over matters both temporal and spiritual. I am sure that neither of us would wish for a repeat of the Venetian situation, so close to the Papal States. Florence is a truly beautiful city, thanks in no small part to the magnificent generosity of its Grand Dukes, and it would be a tragedy to watch its commercial, artistic, financial and diplomatic connections wither and die under a general order of interdiction.

I look forward to seeing you the next time you are in Rome. I still remember our last party – what was that magnificent wine that you had brought with you? I should certainly order a few cases!

Yours amicably, Cardinal Francesco Barberini

Notes on Chapter Nineteen

Galileo

Fictional characters:

None

Galileo Galilei (another figure known by his first name) is still remarkably controversial. Re-evaluations of his importance, or revelations of the secret motivations behind his famous trial, appear regularly and he has been the subject of much writing. Conspiracy theories can be seductive, but the published content of his works was more than enough to prompt action from the ecclesiastical authorities. We can only wonder how things might have proceeded if Bellarmine had survived to preside at the crucial moment.

He was undoubtedly a genius, who made huge contributions to many fields and whose development (if not invention) of the telescope changed the rules of astronomy completely. It would be so much simpler to see his trial, recantation and house arrest as a story of the Church *versus* Science if he had based his arguments on valid scientific reasoning. As it is, his theory of the tides was clearly wrong. It is maddening to the modern reader to find that Galileo explicitly rejected Kepler's theory of the tides (which happened to be correct) and spurned his offer of help. It is also unbelievable that through the whole publication process, Galileo did not realise how provocative the name chosen for the defender of the Ptolemaic system would be. We might conclude that he wished to force the issue, to get the book talked about (which it certainly was, at all levels of society).

The legend that he qualified his recantation with a whispered "*Eppur si muove*" ("And yet... it does move") is precisely that, a legend.

20

Truth

AD 1751

In which we conclude our selection of great astronomical figures and ideas with the greatest of them all, Sir Isaac Newton, who finally teased out the connections between the planets and the Earth, combining superlative scholarship with a pig-headedness even greater than Galileo's.

"You'll have to come closer, you know. My hearing is awful these days. Here, pull your chair over."

Robert wondered, again, whether he was on a fool's errand. He had seen the old man around Cambridge for years. They had even sat on a few of the same committees before all that nonsense about heresy and Whiston's rapid departure from the Anglican Church, but had never spoken other than on matters of business. Perhaps he should just make his apologies and leave. Talking to this fellow would simply be an exercise in frustration. On the other hand, being brutally honest, Whiston wasn't long for this world and it would be a nice gesture to involve him with the project. There were very few people left who had actually known the great man, and failing to consult him would have been an obvious snub. He raised himself a little, and pulled the heavy chair, with its overly ornate gilded arms, a little forward. Very out of date, but in keeping with the rest of the house. It seems that Whiston hadn't re-decorated these thirty years, but then he was scarcely ever in town.

"Is that better, Professor Whiston?"

His voice was pitched half-way between dinner-party loud and hailing a friend across a street, and he wasn't sure how long he could keep going like that. He also wasn't clear whether that was the correct title since all the reshuffling, but knew that correct form was important to a lot of these retired academics. Robert himself liked to think that he didn't particularly stand on ceremony (Plumian Professor of Astronomy, Master of Trinity College – just jobs like any other, if with better perquisites and food), but he had learned from long experience to fit in with the social expectations of others, especially when asking them for a favour.

"Yes, yes, much better. Just call me William, if this is a social visit. May I address you as Robert, Professor Smith?"

Robert nodded, pleased at the simple solution. He resolved to nod or shake his head as much as possible, to get through this ordeal by pantomime. He'd never been very comfortable around elderly people, though heaven knew that most of his colleagues were getting on in years. He wasn't getting any younger himself; perhaps it was good to get used to the idea that he'd probably be like Whiston here in twenty years. Although, unlike this nearly-deaf octogenarian, without a family and a decent place in the country.

"Your letter said something about Newton. Not *another* biography, I hope? I told the last couple of chaps pretty much everything, but they didn't want to hear it. People these days are full of the idea that he was some kind of demi-god, even after thirty years, don't want a word said against him. I think it's national pride, myself, which would have appealed to his vanity. He didn't know everything, anyway – his Biblical work is all just stuff and nonsense, you know, and as for his ridiculous belief in alchemy... So, what d'you want? Please, not another biography."

Robert gritted his teeth, smiling somewhat uneasily. 'Stuff and nonsense' was a fine phrase to use, coming from a man who had scared half of England to death with his predictions that a comet was going to hit the Earth. Nobody could ever accuse Whiston of keeping quiet about his current obsessions. Of course retired mathematicians could be allowed their hobby-horses, but some were less likely to rear up and throw their rider than others.

"No, William. I'm thinking of putting up a monument to Newton. It seems wrong to me that there's one in Westminster Abbey, but not here where he lived, studied, worked."

He looked for a reaction, but Whiston was obviously waiting for him to continue.

"I thought that it should be something done by the college, rather than the university or the city. We could put it in the ante-chapel."

The old man's face suddenly lit up. He barked out an unexpected laugh, eyes sparkling.

"Oh, that's a good one! In the ante-chapel at Trinity, a monument to a man who didn't believe in the Trinity and who was certainly ANTI Chapel. Ha! Anti Chapel, d'you see? The man – oh, excuse me, it's too rich – had to beg for a Royal Exemption from the religious requirements of his post, and you're putting up a statue in a church!"

There was an uncomfortable silence. Robert had thought it rather a good idea. Newton was a fine addition to a long list of Great Trinity Men, and putting it somewhere so near the centre of college life seemed entirely appropriate to him. Perhaps he should go now.

"And you've come to ask me for a donation, I suppose?"

Whiston leaned forward, gazing hard at his visitor.

Robert shook his head violently, wondering whether he was going to be thrown out before he could even explain. Just so long as nobody saw it happening!

"You knew Newton, you worked with him. I was hoping that you could advise about the statue, maybe suggest an inscription?"

"Is this a definite plan, or merely a possibility?"

Thoughts of the fortnight that Robert had spent in London, staying in a cheap lodging-house, reviewing modern funerary statuary flooded his mind. It was hardly his first choice of Long Vacation activity. Of the letters he had exchanged with leading portraitists – much more his field, Trinity was always commissioning paintings. Of the slightly fraught interview that Reynolds had arranged for him with Roubiliac, the talented young Frenchman, and the discussion of terms. So much work – but it was worth it for the sake of the college, and if people in years to come remembered Robert Smith as the Master responsible for that fine tribute to a loyal Trinity man, well, so much the better.

"A definite plan, sir."

That felt much more comfortable than 'William'. He was determined to pay social respect, even if he was fast losing sympathy with this cantankerous survival from an earlier age.

Whiston shook his head.

"Worked with him, you say. Nobody worked *with* Sir Isaac, you worked *for* him. If you didn't work for him, you worked against him, d'you see? I was little better than his secretary, he hated writing letters and wouldn't let me see any of his work until it was ready for publication. Honestly, I've never met anyone quite so fearful, cautious and suspicious that people would steal his ideas. He wouldn't even talk to other college fellows, kept himself locked in his room, had his meals delivered to his door. He neglected the spirit if not the letter of his duties, lecturing to an empty room once he'd driven the students away. The man was a genius, yes, at least in his own area, but he was also a nightmare to work for."

This was a little uncomfortable. Robert had encountered Newton's great mathematical works as an undergraduate. There were several fine copies in the college's new library. However, he had never really cared much to find out about the man. That was unimportant. The work, the theories, that brilliantly incisive mind casting light into the darkness, that was what counted. Not the simply human agent through which that Divine light had been channelled. Maybe it was best to humour the old chap, play along. It would be a miracle if anything came of the meeting.

"I see, sir. How do you think Newton should appear in the statue? I mean, sir, what pose, what expression?"

"A scowl. No, I'm not serious – although he always was. Serious, that is. And scowling most of the time. I can see that you'd have to tone this down if it's going to be half-way decent memorial art. A frown, at any rate, certainly nothing relaxed or suggesting calm. He was never happy, you know, and always puzzled. I don't know which puzzled him more, the universe or the behaviour of human beings."

The old man leaned back in his chair.

"Is it going to be at least half-way decent, d'you think?"

"I should hope so, sir, there's this French fellow who..."

Another unexpected bark of laughter.

"French! Perfect. He hated foreigners on general principle."

Good humour, thought Robert, let it all wash over you.

"He's certainly talented, he's done a statue of Handel that stands in Vauxhall Gardens, and..."

"...a Frenchman sculpts a German! For a pleasure park! My dear fellow, that's perfect, I could think of few combinations that would irritate Isaac more. Certainly a frown, if a scowl is a step too far."

Nobly resisting the urge to point out that both the 'Frenchman' and the 'German' involved had obviously seen the error of their parents' ways and moved to London, the Master of Trinity struggled on.

"Yes, I see. Any other —"

"Hmm, I think that the best pose would be to have him striding forward, walking over everyone in his path. I suppose you couldn't have Hooke, Leibniz and Flamsteed crushed under his unfashionable boot? No, no, I can see that would be too difficult. Anyway, people would argue that Leibniz got the last laugh, as everyone now believes that he invented the calculus. Newton tried hard to stop the truth coming out. Did you know he even stuffed an official committee with his own placeman, and their report reads an awful lot like his style?"

Whiston leaned back in his chair, steepling his fingers and half-closing his eyes.

"And Lord, how we laughed when we heard about Flamsteed's revenge. You don't see that sort of thing happening these days, do you?"

Even Robert had heard about that episode, although the details were hazy. He had been, what, in his early twenties when old Davidson, who shared his lodgings, had come back from London full of the doings of the Royal Society members, quite star-struck. How the President of the Royal Society and the Astronomer Royal had fallen out to the extent of the latter making a bonfire of the former's books in front of the Greenwich Observatory.

"What exactly happened then, sir?"

"It was Newton's famous defeat, the only time anyone stood up to the bully. Glorious, Robert, glorious! Sir Isaac had stolen figures – yes, that's the only word for it, stolen – from Flamsteed's unpublished work, d'you see, so he bought up every single copy of the unauthorised book and burned 'em all! Newton locked himself in a room for a week, wouldn't even see Halley. Now, *there* was someone who was a scholar *and* a gentleman."

He closed his eyes, smiling at the recollections.

"Poor Halley. Poor Flamsteed. Never got the credit they deserve. Isn't there a line in Lucretius somewhere, the rising sun drowns out all the other stars? Of course, Newton was quite, quite, brilliant. In his own fields, mind."

Robert made a mental note – at least the serious expression and the idea of moving forward, if not over the reputations of his rivals, were good ideas, that he would pass on to Roubiliac when they next met. That was as much as he could hope for from the afternoon, other than the surprisingly good cup of coffee he'd been served while waiting. He'd have to find out where they got the beans from. Still, it seemed a bit surly to break off conversation after only ten minutes.

"What would you say his field was, sir? If you had to come up with an epigram for the memorial, something short and memorable?"

"Well, where to begin? Mathematics, astronomy, optics, mechanics – you might as well say all of physical nature. Our ancestors split the world into the physical, the Divine and the human, did they not? So, fully one-third of knowledge. I suppose it wasn't his fault that he thought he should tackle two-thirds, but trust me, Robert, he was no theologian. His half-baked Biblical ideas had really no basis, really no basis at all. He stumbled on some important theological truths, him and Locke, but stumbling was all that they were doing. I could tell you about the time—"

It all flowed over Smith. He'd never really seen the point of the academic study of theology. Newton had been his inspiration in the physical sciences, though, and at least a part of his own studies in optics had been inspired by his great Trinity predecessor. What a satisfying blend of experiment and mathematics could be seen in the precise curve of a lens! Surely, that was Newton's greatest legacy, the marriage of physics, astronomy and mathematics. For years, centuries, millennia people had stumbled (good word, that) through the universe, slicing the cake in a way that was most unpalatable. What happened on Earth had no relevance to what happened in the Heavens, and the other way around. Separate maths for separate problems. A few people had approached the truth – Kepler and Galileo, a scant century ago – but could not make that final step, the one that needed a solid English genius. A Cambridge genius. A Trinity genius! Sir Isaac Newton had re-invented heaven and Earth and shown that they were the same, that the same maths applied, the same laws, the same ideas. Laws and ideas that followed a precise, formal logic, that admitted no interference either human or Divine... a fresh start for humanity. Yes, he wasn't a nice man to know by all accounts, and yes, he wasted much of his time and energy on occult theology, alchemy and the like. But still, thought Smith, the greatest thinker of our age. Of any age?

"— could you?"

He started, and suddenly became aware that Whiston was awaiting a response.

"I'm sorry, sir?"

"You couldn't use those lines by Pope, I said. It would be a bit below the dignity of the Master of Trinity to commission a sculpture of his hero and then finish it off with some doggerel by a modern comic poet!"

"Oh, absolutely. It has to be something classical, that can sum up just how out of the ordinary the man was. I'd prefer a genuine tag rather than getting someone to write anything new."

"You know, I've remembered that Lucretius passage. You have read *De Rerum Natura,* haven't you? It's an interesting take on this atomism question that got so vexed recently. He's talking about some philosopher or another, and he calls him *qui genus humanum ingenio superavit et omnes perstrinxit stellas exortus ut aetherius sol.*

Who surpassed the race of men in understanding as a rising sun obscures the stars, d'you see? I'm fairly sure that he was talking about Epicurus. It's a fairly famous line, I'm sure one of your colleagues could check it for you."

That's it! How perfect! thought Robert. No need for a first name, or title, or dates – he's a man for the ages, for all generations. Just "Newton: Qui superavit..." or whatever it was. He smiled broadly, the first genuine smile of the afternoon. What a success the visit had been! How glad he was that he had come!

"Why, sir, I think that is very dignified, and very appropriate. There's certainly no need to mention the lesser stars whose light was over-flooded by Newton. Thank you so much for your time, your patience with me, sir, and may I wish you a pleasant stay in Cambridge and a good journey back to... Rutland."

Was it Rutland? He stood, extending his hand, his mood buoyant.

William Whiston, formerly Lucasian Professor, leaned forward to clasp the hand of his excited visitor. What was that Shakespeare line about the mind's construction being visible in the face? His memory really wasn't what it was. Anyway, this Robert Smith had been very easy to read, his irritation, boredom and excitement like those masks the Greek actors wore.

"Goodbye, my dear fellow. I'll come to see your statue, if I'm spared."

He released hands, smiling as the portly Smith bowed his way to the door and clattered down the stairs. He leaned back in his chair, smiling to himself. Newton as Epicurus – it was Epicurus, wasn't it? How fitting! One should seek neither pleasure nor pain but the absence of both. The gods exist and should be remembered, but they have no interest in the affairs of mortal beings. How fitting! He'd never asked Newton his opinion on the immortality of the soul, but he might well have agreed with the philosopher who had supposedly said, "I was not; I have been; I am not; I do not care." He'd never asked Newton anything more interesting than how many copies to make, or what he should fetch for lunch, come to that.

Newton, Newton, what are we to make of you?

He asked himself, as he settled for his afternoon nap. He rang the bell that would summon his servant with his favourite blanket, the one with a Greek key pattern.

Were you, after all, the end of the story? Or the beginning of a new one? It's for others to find out, I've done my bit. I wonder what new theories will be on sale in the grove of the Academy a hundred years from now?

New stars for sale, new stars for old.

Notes on Chapter Twenty
Newton

Fictional characters:
Davidson

Sir Isaac Newton was a difficult person to deal with, arrogant and secretive. Everything said about him in the chapter is true. He spent much of his time alone in his rooms, attempting to decode the hidden messages in the Bible, or working on alchemy. In his professional life, he would tolerate no rivals and insisted on being recognised as pre-eminent. The chapter does not provide an exhaustive list of his feuds: for example, he refused to join that élite organisation of British scientists, the Royal Society, until Robert Hooke had died, so that they would not encounter each other at meetings.

The framing story about his statue uses real characters, whose biographical details are correct, but their personalities have been completely invented. The monument itself is very fine and well worth seeing when in Cambridge.

Ask any scientist to name the two greatest physicists of all time, and the odds are good that the answer will be Newton and Einstein. Newton discovered nearly all of the physics that one learns at school, and transformed our understanding of the universe by demonstrating that the rules which governed the motion of objects in the heavens are *the same* as those governing motion on Earth. Astronomy had become a branch of physics, which is a fitting place to end this sequence of stories.

Afterword

In the period up until Newton there were two great questions for those who studied the night sky. Astronomers wondered, "how exactly do the planets move?" Cosmologists asked, "what makes them move like that?"

The theories introduced in the stories all had their day, and it seems right to bring them all together in one place to round off my samples from astronomical history. Some things have been beyond doubt throughout the whole period covered by the book; for example, from very ancient times we've known that the Earth is roughly spherical, and that the Sun is the only object in the Solar System giving off its own light.

As an aside, in ancient and medieval authors, the word "mathematics" often means either "astronomy" or "astrology" – and "world" often means "the whole universe" rather than just the Earth. I make no apologies for having stuck to the modern meanings!

Aristotle, no great astronomer himself, was hugely influential throughout history. His books, and the many commentaries upon them, became the first reference-point for almost any knowledge (a one-man Wikipedia). He adopted a theory proposed by his colleagues Eudoxus and Callipus; the Earth sits in the middle of the universe, because earthly elements naturally fall to the centre (which is why stones fall downwards). A vast machinery of spheres surrounds it, with each sphere rotating at a slightly different speed. The planets are stuck to some of these spheres. It's very likely that he viewed the spheres as real. They just happen to be invisible, as they are made from an element we don't find on Earth, *aether*.

Unfortunately, Aristotle's spheres didn't get things quite right. For a start, they don't explain why the Moon gets larger and smaller. The obvious answer is that it's changing its distance from the Earth, but if it's stuck to a sphere it couldn't do that.

Claudius Ptolemy, the greatest figure of ancient astronomy, spent his entire life devising a magnificently complicated, Heath Robinson, system. Here, each planet was attached to a small circle (an *epicycle*). The centre of this epicycle (the *epicentre*) itself moved around a large circle; but this large circle wasn't *quite* centred on the Earth, it was off-set a little bit (to make it an *eccentric*). Finally, the epicentre didn't move evenly around the eccentric; it changed its speed as it went.

Essentially, Ptolemy had allowed himself so many variables to play with that, by tweaking things a little here and there, he could perfectly re-create the observed movements of the planets in the night sky. It was a triumph of dogged mathematical determination; but it still got things wrong for the Moon, which now should have changed its size even *more* than is actually observed. However, nobody seemed to notice this for well over a thousand years.

It's still an open question for historians whether Ptolemy thought that his circles actually existed in the heavens, or whether they were neat mathematical tricks. By the time that the mystical philosophy of Neo-Platonism arose, with such figures as Proclus, there was little doubt that astronomers were thoroughly split on this question.

After the general collapse of intellectual culture in Western Europe, astronomy was kept alive by Muslim scholars and philosophers, across the Islamic world. They wholeheartedly adopted Ptolemy's system, introducing a huge number of improvements and mathematical innovations; and by the time that the knowledge started to come back to the remains of the Roman Empire, technical astronomy was pretty much fixed into a form that would last for hundreds more years.

The debate shifted to questions about the interpretation of these mathematical rules, and the relations between astronomy, astrology, philosophy, and religion. The rise of the university system encouraged freedom of thought, leading to clashes with all forms of authority. Then, as Europe entered its Renaissance, science of all kinds found itself at the forefront of ideas; providing a chance for modern thinkers to show themselves better than their ancient forebears.

So when Copernicus proposed that the Earth was a planet, which moved around the Sun, the scientific world was ready for something new. The theory was a little more complicated than a simple swap. Copernicus wanted the Earth to move in a perfect circle, with no eccentrics or epicycles, so to get things right he had to say that the Sun moved a little bit. For him, this was no mathematical fiction. It was a true picture of the universe, drawn to glorify the Creator. If Copernicus saw, on his deathbed, the preface attached to his book saying that it wasn't to be taken seriously, it probably finished him off.

The Copernican theory rapidly supplanted the Ptolemaic theory throughout Europe. Some careful astronomers such as Tycho Brahe were not convinced. A moving Earth should give rise to changing patterns in the fixed stars. Either the stars are an immense distance away (which Tycho ruled out as ridiculous) or the Earth is stationary. He came up with a short-lived compromise, in which the planets orbit the Sun – but the Sun orbits the Earth.

Tycho's longer-lasting contribution to cosmology was to shatter the heavenly spheres once and for all. His detailed study of a comet showed that it started from, and returned to, the distant space beyond the planets, passing them as it first approached and then left the Earth. Despite some half-hearted resistance, Aristotle's idea was finally discredited.

Kepler, like Ptolemy so many years before him, dedicated his life to patient mathematical analysis of the planetary movements (inspired by similar religious motivations). Gone were combinations of circles; he proposed that heavenly bodies, including the Earth, move in elliptical orbits. He also showed that the question "what's in the centre?" is an empty one; how the planets appear to move in the night sky depends on where we are standing. We only think that the Earth is special because it's our natural point of reference.

Galileo mopped up many of the objections that non-scientists had to the Sun-centred model. He used his telescopes to show that other planets were similar to the Earth in various respects; they had mountains, and had moons of their own.

Neither Kepler nor Galileo could explain convincingly *why* the planets moved as they do (although Kepler had a decent attempt at using magnetism as a starting-point). It took Newton to show that the same force of gravity which causes an apple to fall on Earth also causes a planet to orbit the Sun. What's more, the planets obey exactly the same mathematical equations as do objects on Earth. Since the very start of Greek astronomy, everyone had assumed that heavenly movements would follow different laws of physics from Earthly ones, so Newton's insight completely changed the rules of the game.

All those who followed him would agree that astronomy was now a part of physics, rather than a separate subject in its own right, neatly wrapping up another ancient debate. Standing at the end of this sequence of stories, Newton marks the beginning of our modern conception of science; but that's a story for another day.

Acknowledgements

It takes a lot of people to make a book, not just an author.

Writing this was only possible with the support of my wife, Jenny, and my children, Katy and Dan. All of those early mornings and weekends spent tapping away at the computer only happened through their patience and understanding.

I was inspired and encouraged to study the history and philosophy of science by some great teachers; Peter Fekete (Victoria College, Jersey), Bill Newton-Smith (Balliol College, Oxford), the late Peter Lipton (Department of HPS, Cambridge), and John Milton (King's College, London).

These stories have been going around in my head for years, and I've been fortunate enough to share them as part of my job. I taught a "History and Philosophy of Science and Society" class at Edmund Burke School in Washington DC, and I've also enjoyed discussing these characters and ideas with staff and students at the Royal Grammar School, Newcastle-upon-Tyne.

My friends have provided honest feedback about drafts and excellent conversation, often accompanied by good food and wine – especially Lisa Crawford, Alex Norrish-Kleine, Tora Smulders-Srinivasan, Sarah Mackie, Chris Petkov, Mark Poles, Pete Clark, and Anna Bowles.

You wouldn't be reading this at all if it weren't for the staff of Candy Jar Books, and I hugely appreciate the work done by my publisher, Shaun Russell, and my editor, Rebecca Lloyd James.

Newcastle-upon-Tyne, 2013